Digital Twins and Simulation Technology

This book provides a comprehensive overview of the concept of digital twins, emphasizing its strategic importance across various commercial domains. This book covers the fundamentals, data requirements, tools, and technologies essential for understanding and implementing digital twins. It discusses how digital twins are used for running simulations, analyzing performance issues, and generating potential improvements to optimize business processes. The book explores the architecture, historical background, and real-time applications in sectors including urban planning, healthcare, smart cities, and manufacturing.

- Explains digital twin technology, including its core principles, architecture, and how it replicates physical objects in virtual platforms, in detail
- Covers the data types and tools necessary for creating and maintaining digital twins, including sensors, data processing systems, and integration methodologies
- Explores enabling technologies such as Computer Vision, IoT, AI, ML, 5G, AR, and VR that drive the functionality and application of digital twins
- Analyzes practical applications in diverse sectors like urban planning, smart cities, healthcare, manufacturing operations, and power-generation equipment, showcasing real-world use cases and benefits
- Examines real-time challenges and limitations associated with implementing digital twin technology, providing a balanced view of its capabilities and constraints

It is a reference book for researchers, scholars, and students who are working or interested in learning about digital twin technology.

Digital Twins and Simulation Technology

Concepts and Applications

Edited by
Lavanya Sharma and Pradeep K. Garg

CRC Press
Taylor & Francis Group
Boca Raton London New York

CRC Press is an imprint of the
Taylor & Francis Group, an **informa** business
A CHAPMAN & HALL BOOK

Designed cover image: ShutterStock

First edition published 2026
by CRC Press
2385 NW Executive Center Drive, Suite 320, Boca Raton FL 33431

and by CRC Press
4 Park Square, Milton Park, Abingdon, Oxon, OX14 4RN

CRC Press is an imprint of Taylor & Francis Group, LLC

ISBN: 9781032949390 (hbk)
ISBN: 9781032949413 (pbk)
ISBN: 9781003582489 (ebk)

DOI: 10.1201/9781003582489

Typeset in Times
by Deanta Global Publishing Services, Chennai, India

Dedicated to my Dada Ji (Late. Shri Ram Krishan Choudhary Ji) &
In memory of Ladoo, my truest companion and quietest inspiration

– Dr. Lavanya Sharma

Dedicated to my Parents (Late Shri Ramgopal Garg and Late Smt Urmila Garg)

– Prof. Pradeep K. Garg

Contents

Preface ix
Editor Bio x
List of Contributors xii

Chapter 1 An Introduction to Digital Twins and Simulation 1
Lavanya Sharma, Pradeep K. Garg, and Nitin Tripathi

Chapter 2 A Multilayered Architecture of Digital Twin Technology 13
Naba Suroor

Chapter 3 Understanding Data Framework for Digital Twinning 27
Rajeev Kumar, Balaji Aryan Singh, and Mrityunjay Sharma

Chapter 4 A Digital Twinning for Advanced Precision Medicine and Patient Care: Tools and Technologies 40
Sudhriti Sengupta, Lavanya Sharma, Abhishek Kumar, and S B Goyal

Chapter 5 Metaverse Marketing Strategies Using IoT and 5G 55
Subhranil Som and Asik Rahaman Jamader

Chapter 6 Geospatial Digital Twins: An Emerging Technology in GIS 74
Pradeep K. Garg and Martin Kappas

Chapter 7 A Cardiac Medicine and Surgery Using Digital Twinning: A New Era in Patient Care 85
Lavanya Sharma, Ankit Garg, and Mukesh Carpenter

Chapter 8 Applications of Digital Twins in Personalized Medicine and Healthcare 95
V. Jokanović and M. Živković

Chapter 9 Building Future-Ready Cities: The Power of Digital Twins in Indo-Dutch Collaboration 116

Sudesh Agrawal, Naman Agrawal, Himanshu Agrawal, and Sonal Agrawal

Chapter 10 A Transductive and Inductive GNNs for Physical Moving Objects Detection in Surface Scenes for Digital Twins 133

Wieke Prummel, Jhony Giraldo, Badri N. Subudhi, Anastasia Zakharova, and Thierry Bouwmans

Chapter 11 Toward Graph Refactoring in Latent Space for Underwater Moving Object Detection for Digital Twins 150

Meghna Kapoor, Badri Narayan Subudhi, Ankur Bansal, and Thierry Bouwmans

Chapter 12 Toward a Complete System for Pupil Monitoring Using Deep Learning for Digital Twins 167

Kamal Sehairi, Patrick Franco, El-Hadi Zahzah, and Thierry Bouwmans

Chapter 13 Emerging Application of Digital Twin Technology toward Development of Smart Infrastructure 186

Jimmy Mehta, Yaman Hooda, and Yati Aggarwal

Chapter 14 Navigating the Complexities: Limitations and Challenges of Digital Twin Technology in Smart Cities 201

Sudesh Agrawal, Naman Agrawal, Himanshu Agrawal, and Sonal Agrawal

Preface

This book explores the utilization of digital twin technology for next-generation healthcare, urban planning, industries, and infrastructure using a series of present tools of data collection and applications. It covers the transformational impact of digital twins, from their origin in engineering to their real-time applications in manufacturing, healthcare, urban planning, power, and other areas. It will present real-world case studies focusing on real-world outcomes, while potential risks are addressed wisely. It covers the challenging issues related to twin technology and solutions to handle relevant challenges.

This book comprises four parts that provide an overview of basic concepts from historical background and architecture twin technology with IoT, computer vision, artificial technology, augmented reality, virtual reality critical applications domains, tools, technologies, challenges, and solutions to handle the relevant challenges in smart cities including healthcare, manufacturing industries, and many more. This book provides a detailed description to the readers with practical ideas of twin technology to deal with human dynamics, challenges involved in surpassing diversified architecture, communications, integrity, and security aspects. Digital twin technology in combination with other technology proved to be most advantageous for the companies to efficiently monitor and control their day-to-day process.

Overall, the publication of *Digital Twins and Simulation Technology: Concepts and Applications*, would help the readers understand in-depth knowledge of digital twin technology to students, engineers, professionals, and researchers, as well as organizations.

Editor Bio

Lavanya Sharma received her PhD (Computer Science) from Uttarakhand Technical University (State Government University), India, in 2018, in Digital Image Processing and Computer Vision. She also received *TEQIP* scholarship for the same. She received her MTech (Computer Science) in 2013 from Manav Rachna College of Engineering, affiliated to Maharshi Dayanand University, Haryana, India. She is currently working in Department of Information Technology at New Delhi Istitute of Management as an Assistant Professor and Senior team member in Research and Innovation department also. Her work is on Motion-based Object Detection using Background Subtraction Technique for Smart Video Surveillance. She is recipient of several prestigious awards during her academic career and has qualified certification courses from *ISRO Dehradun Unit, India.* She has published eight books (authored and edited) and over 70 research articles. She has four patents in Computer Vision (Object Detection and Tracking). She has acted as Organizing Committee Member, Conference Chair, Session Chair, Reviewer, and Technical Program Committee (TCP) member for various national and international conferences. She has developed various new courses and practical exercises in Computer Science Engineering. Besides supervising a large number of undergraduate projects, she has guided more than 40 postgraduate projects. Presently, she is Editorial Member/Reviewer of various journals of repute and an active program committee member of various conferences. Her primary research interests are digital image processing and computer vision (detection of foreground object in challenging surface), artificial intelligence, machine learning, deep learning,remote sensing, GIS and Internet of Things (Google H index-20, I Index 34).

Pradeep K. Garg worked as Vice Chancellor at Uttarakhand Technical University, Dehradun, India. Presently he is working as Professor in the Department of Civil Engineering, IIT Roorkee, Uttarakhand, India. He completed BTech (Civil Engg.) in 1980 and MTech (Civil Engg.) in 1982 both from the University of Roorkee (now IIT Roorkee), Uttarakhand, India. He is a recipient of Gold Medal at IIT Roorkee to stand first during MTech program, Commonwealth Scholarship Award for doing PhD from the University of Bristol (UK), and Commonwealth Fellowship Award to carry out postdoctoral research work at the University of Reading (UK). Prof. Garg has published more than 300 technical papers in national and international conferences and journals. He has undertaken 26 research projects and provided technical services to 83 consultancy projects on various aspects of Civil Engineering, generating funds for the Institute. He has authored and edited various books and produced two technical films on Story of Mapping. He has developed several new courses and practical exercises in Geomatics Engineering. Besides supervising a large number of undergraduate projects, he has guided about 74 MTech and 30 PhD theses. He is instrumental in prestigious MHRD-funded projects on e-learning; Development of Virtual Labs, Pedagogy, and courses under NPTEL. He has served as an expert on various national committees, including the Ministry of Environment & Forest, EAEC Committee, NBA (AICTE), and Project Evaluation Committee, DST, New Delhi. Prof. Garg has reviewed a large number of papers for national and international journals. Considering the need to train human resources in the country, he has successfully organized 44 programs in advanced areas of Surveying, Photogrammetry, Remote Sensing, GIS, and GPS. He has successfully organized 10 conferences and workshops. He is a life member of 24 professional societies, out of which he is a Fellow member of eight societies.

Contributors

Pavle Aćimović
Princeton University
Princeton, New Jersey

Himanshu Agarwal
3rdTech
India

Naman Agarwal
Embassy of the Kingdom of Netherlands for India
New Delhi, India

Sonal Agarwal
Krishi Neer
India

Sudesh Agarwal
Cleveland Clinic
Ohio

Yati Aggarwal
Manav Rachna International Institute of Research and Studies
Haryana, India

Ankur Bansal
Indian Institution of Technology Jammu
Jammu, India

Thierry Bouwmans
Lab. MIA
La Rochelle Université
La Rochelle, France

Mukesh Carpenter
Alshifa Multispecialty Hospital
Delhi, India

Ankit Garg
Department of Paediatric Cardiology
Amrita Hospital
Faridabad, India

Pradeep K. Garg
Civil Engineering Department
Indian Institute of Technology Roorkee
Uttarakhand, India

Jhony Giraldo
Télécom Paris
Paris, France

S. B. Goyal
Faculty of Information Technology
City University
Malaysia

Yaman Hooda
Department of Civil Engineering
Manav Rachna International Institute of Research and Studies
Haryana, India

Asik Rahaman Jamader
Pailan College of Management and Technology
West Bengal, India

V. Jokanović
Innovative Company ALBOS doo
Belgrade, Serbia
and
Institute of Nuclear Sciences "Vinča"
Belgrade, Serbia

Martin Kappas
Cartography, GIS & Remote Sensing Department, Institute of Geography
Georg-August University of Göttingen
Germany

Meghna Kapoor
Indian Institution of Technology Jammu
Jammu, India

Abhishek Kumar
Department of General Surgery
NSMCH
Patna, India

Rajeev Kumar
Centre for Innovation and Incubation
Bhubaneshwar, Odisha, India

Jimmy Mehta
Department of Mechanical Engineering
Manav Rachna International Institute of Research and Studies
Haryana, India

Patrick Franco
Lab. L3I
La Rochelle Université
La Rochelle, France

Wieke Prummel
Lab. MIA
La Rochelle Université
La Rochelle, France

Lavanya Sharma
Department of Information Technology
New Delhi Institute of Management
Delhi, India

Mrityunjay Sharma
UCL School of Management
London, UK

Badri N. Subudhi
Indian Institute of Technology Jammu
Jammu, India

Kamal Sehairi
Laboratory of Applied Sciences and Didactic
Ecole Normale Supérieure de Laghouat
Laghouat, Algeria

Balaji Aryan Singh
Department of Computer Science, Institute of Technical Education & Research (ITER)
SOA University
Bhubaneswar, Odisha, India

Naba Suroor
NITI Aayog
Delhi, India

Sudhriti Sengupta
Amity Institute of Information Technology
Amity University
Uttar Pradesh, India

Subhranil Som
Bhairab Ganguly College
West Bengal, India

Nitin Tripathi
Asian Institute of Technology
Klong Luang, Pathumthani, Thailand

El-Hadi Zahzah
Lab. L3I
La Rochelle Université
La Rochelle, France

Anastasia Zakharova
Lab. MIA
La Rochelle Université
La Rochelle, France

M. Živković
School of Dentistry
Belgrade, Serbia

1 An Introduction to Digital Twins and Simulation

Lavanya Sharma, Pradeep K. Garg, and Nitin Tripathi

1.1 INTRODUCTION

Digital twin (DT) is an innovative encompassing tool that has various advantages such as real-time monitoring, optimization, simulation, and real-time forecasting and can be used in manufacturing industries and healthcare. However, the framework and practical implementation of this technology are yet to be fully achieved for this concept at scale. Presently, several organizations have implemented this technology but no public data is available related to its implementation, which makes it difficult to access their components and effectiveness effectively. The advancement of machine learning (ML), Internet of Things (IoT), artificial intelligence (AI), data analytics, and big data has shown a notable forwarding in real-time monitoring and forecasting [1–3]. But due to its several open challenges, there exists a gap in this domain that has prevented the adoption of this technology. On the basis of recent and past work done by researchers in industry, a summarized view of DT is presented in this chapter, which includes simulation, share market scenario, types, advantages, and disadvantages of this emerging technology [4–9].

In today's world, simulation is a modeling tool that is used to create infrastructure to monitor the key features of the actual system by transferring data from any physical system to the virtual world. Using a simulation tool, the manufacturing process becomes traceable and benefits in cost, time, and risk management. A Digital Twin (DT) integrates the data, simulation models, and all other relevant information from the physical system into various domains where they are required for designers, operators, or technicians. The core component of DT is simulation, which is used to implement the behavior of the object (simulated) as shown in Figure 1.1. In recent years, simulation and mathematical models have improved DT significantly due to problem-solving and sematic methods and the availability of virtual technology. Most of the companies are working on DT models such as "Siemen," which is building complex technological products. They are creating DT of their product in order to design them in less time, be powerful, and also enable their customers to use

DOI: 10.1201/9781003582489-1

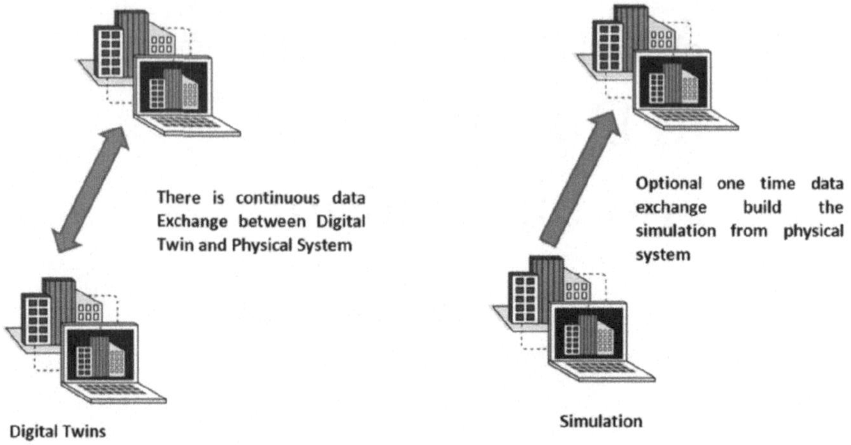

FIGURE 1.1 Working of digital twin and simulation models

them more effectively [1–4, 8]. For example, This Company is looking to make food articles production more sustainable and efficient.

The concept of DT has emerged as a transformational prototype that bridges the physical world to the virtual world in today's dynamic environment where technologies keep on changing. DT is a replica of the real world and may be a system or process or object. Because DT data can be integrated from various sources, the virtual version of this can be used as a powerful tool for detection, tracking, and reproducing the corresponding actions. DT can allow predictive insights, support decision-making, and improve diagnosis by creating a digital replica of a particular object such as cancer detection, heart surgeries, orthopedics, liver treatment, and many more. DT reflects the dynamic reflection of physical objects by gathering data from sensors, simulations, and past records [5–10]. The data scientist then utilizes this integrated data which overall impacts the development of new technologies such as Internet of Things (IoT), artificial intelligence (AI), deep learning (DL), and machine learning (ML). In addition to this, data analyst can be employed in the manufacturing domain [11, 12, 15–28].

This work comprises the significance of simulation in DT, market size in healthcare, pros and cons, and different types of DT. In this chapter, Section 1.1 presents a detailed overview of this innovative technology and its present role in the industry, Section 1.2 deals with simulation and its role in DT. Section 1.3 presents the analysis of global DTs in the healthcare sector's market size, share, and industry trends for 2019–2030. In Section 1.4, various types of DT are discussed, and in Sections 1.5 and 1.6 the advantages and disadvantages of this technology are discussed.

1.2 SIMULATION IN DIGITAL TWINS (DTS)

This innovative technology combines twinning, optimization, simulation, analytics, and real-time monitoring. This technology can be viewed as an upcoming advancement in digitization and simulation as it can save time and cost. In this case, no physical prototype is required; a virtual prototype can be built without affecting the present operations. Simulation plays a vital role in DT as it allows users to check scenarios virtually [12–15] and helps in supporting data-driven decisions such as diagnosis, digital building of machines, customer experience, predictive maintenance, prediction, and optimization as shown in Figure 1.2. Some of them are listed below:

a. **Predictive Analysis:** It employs the system's physical representation to simulate its future behavior and on the basis of simulation prediction can be done. In DT, simulation can analyze the performance of an object or system or product for future perspective. Overall, it helps in decision-making and leads to high adaptability [].
b. **Trial-and-Error Reduction:** It is a fundamental technique of solving a problem by repeated and varied trials till the main goal is achieved or a solution is found. In problem-solving, DT remarkably reduces trial and error in manufacturing and supply chain. Simulation also helps in the reduction of any physical change, which leads to cost and time saving [2–7, 10, 16, 17].
c. **Enhanced Supply Chain Management:** DT and simulation have improved collaboration and decision-making in supply chain operations. They require integration of data and steady enhancement to accurately optimize and monitor physical supply chain operations. Simulation and mathematical models for creating DT is improved significantly [18, 19]. In DT, simulation helps in checking stocks, demand deviations, and inventory scenarios.
d. **Accelerating Development:** This tool is accelerating development in various sectors of industries such as manufacturing, automotive, and healthcare. Simulation helps engineers to design systems that require fewer changes in models and fast development. It takes less time to iterate and test products, which results in less cost and time in the development cycle []. It also helps

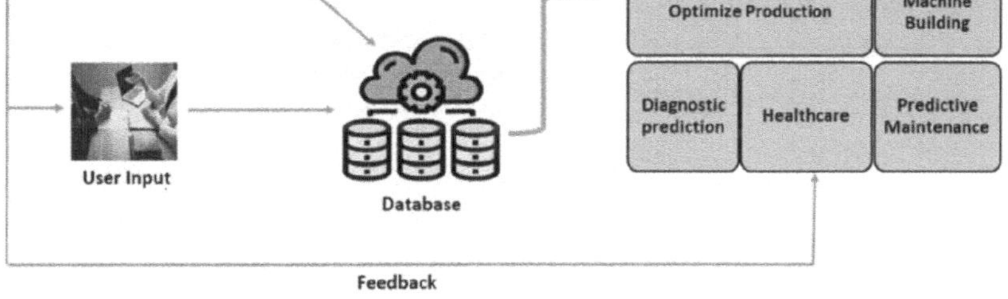

FIGURE 1.2 A digital twin framework

in speedy drug discovery, sustainable energy management, and product quality.

e. **Enhanced Product:** Performance simulation helps to analyze performances and develop new ideologies for products. The simulated system or object can be easily integrated into the digital platform, which represents accurately in an efficient manner. DT and simulation can improve products by refining their product design, manufacturing, and support [23–28].

1.3 ANALYSIS OF DT'S MARKET SIZE AND TRENDS IN HEALTHCARE DOMAIN

The Global DT in the healthcare sector will approximately reach $29.3 billion by 2030 as per the report [2], with the rising global market growth of 51.8% compound annual growth rate (CAGR) during the prediction period. Globally, in the healthcare sector, the surgical planning and medicine market is the main focus of extensive scientific research projects. It registered $81.3 million in revenue in the global market in 2022. With outstanding accuracy, DT tracks and updates physical systems or objects in real time [22, 23, 29–33]. Overall, it led to an increase in scheduling, cost, time, less chances of error and less risk management, and improved supply chain using artificial intelligence (AI), data analytics, and machine learning (ML) as shown in Figure 1.3. Due to prevalent usage and the need for high precision, the global market is expected to increase further in the upcoming years.

**In August 2021, IBM collaborated with Black & Veatch to aid end users keep assets and equipment running at top performance and reliability. In November 2022, SAP made an agreement with Schneider Electric (SE) to enhance digital transformation and innovation to streamline onboarding and assets management with DT, Augmented reality (AR), and Virtual Reality (VR) [21, 33–35]. In April 2023, Oracle collaborated with another firm "Deloitte Health" – Oracle Accelerated. As per the

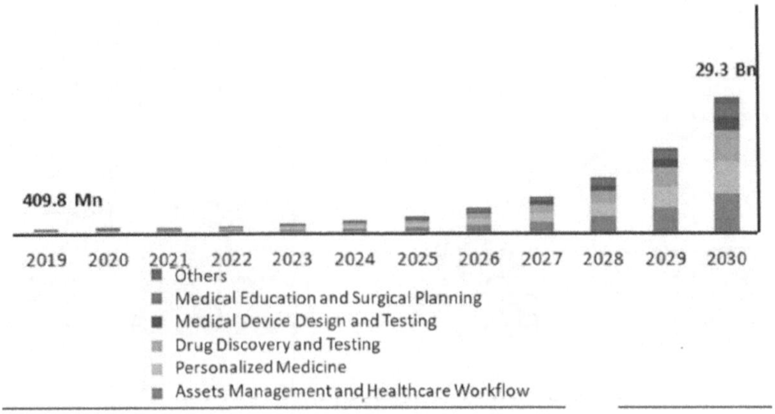

FIGURE 1.3 Analysis of global DTs in healthcare sector: market size, share, and industry trends for 2019–2030

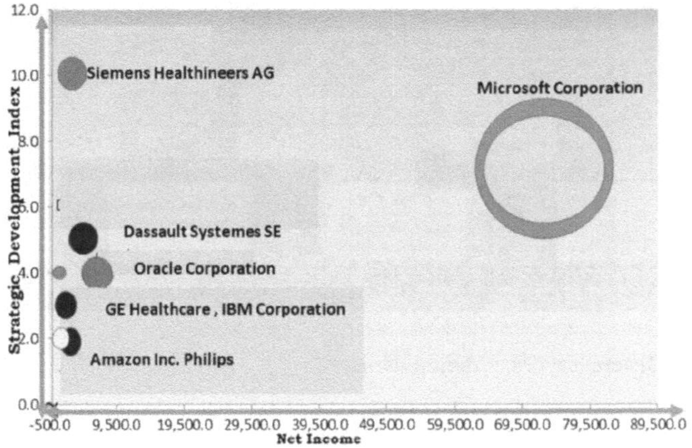

FIGURE 1.4 Market competition analysis by KBV Cardinal Matrix

KBV Cardinal Matrix report, Microsoft Corporation is the forerunner in the market. In 2022, Microsoft tied up with Cognizant to provide new digital health solutions to enhance patient monitoring remotely [36–38]. On the other hand, Siemens AG, Dassault Systems, and Oracle are among key innovators in the present market as shown in Figure 1.4.

1.4 TYPES OF DTS

The use of DT and simulation or mathematical models has become a crucial tool in industries such as healthcare and manufacturing to respond to disruption and complexity. Presently, various manufacturers are working on DT. By creating virtual copies of assets, systems, and processes, they facilitate lower risk, operational improvement, and time and cost reduction [36, 39, 40]. Through the integration of data collected from different resources on a real-time basis, such as sensors and machines, DT provides an effective model of how things behave in real-time scenario. This empowers firms to monitor, design, and optimize their operations more accurately and effectively. DT is a bridge between a physical system and a virtual system. It represents an object or system or physical entity [41–45]. There are four types of DTs, as shown in Figure 1.5, which are listed below:

a. **Component Twins:** It is also known as parts twin, which is the lowest level of DT. It corresponds with the system's smallest element or a particular part of a product such as sensor, valve, switch, etc. The virtual model of these parts provides real-time monitoring regarding their performance to test their stability and accuracy. It is the basic form of DT and focuses on a particular system's part or machine [45–47]. They stimulate the minute functional components to maintain data about its present situation and performance.

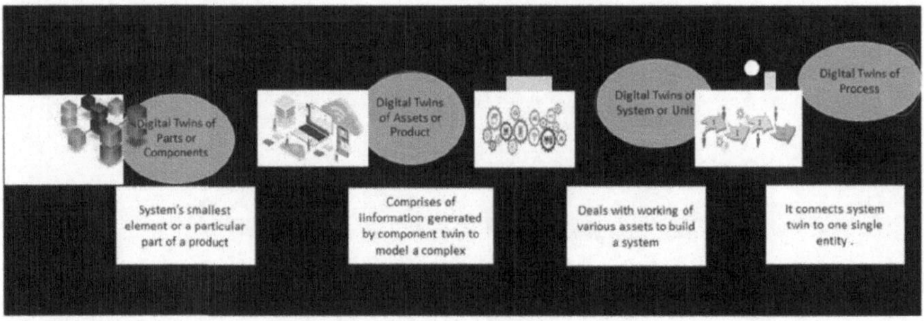

FIGURE 1.5 Different types of digital twins

b. **Asset Twins:** It is also known as Product Twin, and is the second level of DT. This Twin consists of information generated by component twin to model a complex asset such as engine, building, or pump in the case of the manufacturing industry. This twin analyzes the efficiency of each and every part and performs as a complete solution. This stimulates all parts of the system or machine and interaction [45, 48]. It also describe working of particular component constitute a complete assets.
c. **System Twins:** It is also known as Unit Twin, as it replicates assets on the system level, and deals with combinations of all assets into a single functional unit. This provides a large-scale idea of the plant or firm or industry, by testing different system configurations to get new business opportunities for additional profit sources. Overall, this twin provides a detailed image of various correlated assets comprising a system. It also deals with the working of various assets to build a system [49, 50].
d. **Process Twins:** It optimizes system performance and workflow by stimulating the complete process or system. It consists of algorithms that require a team of engineers who are experts in developing the algorithms. This is the highest level among all other twins as it connects the system twin to one single entity that synchronizes between different systems. This helps to adapt inputs such as production temperature, raw input data, and output data without affecting the actual manufacturing process and production quality [51–54]. In this manner, the team of engineers can test different business hypotheses effectively and efficiently.

1.5 DISADVANTAGES IN IMPLEMENTING DT

A DT is basically a virtual replica of a real-time system or an object or procedure. While implementing DT has various advantages, there are also open challenges or disadvantages. For example, a car manufacturing firm can build DT of its organization. So for this purpose, DT must consist of physical view, software, and data that connect them. In this case, also DT has to face various open challenges [3, 10, 16, 23, 34, 50, 54]. Some of them are listed below:

a. **High Complexity:** Organization that are creating DT of new factories, products or process might be easy but if any company with existing Information Technology (IT) infrastructure need to modify or alter the present data in their system before building the DT which will be very cost effective [54, 55].
b. **Customization Requirements:** Organization Engineers must do an analysis before building DT for their organization as per their need and requirements [55, 56]. Some organizations prefer one-size-fits-all solution clause, but in the case of DT, it should be customized as per their goal and requirement, which is time-consuming and less cost-effective too.
c. **High Cost of Implementation and Long-Term Maintenance:** The process of implementing DT is very expensive compared to the calculated predicted cost. The amount associated with building and processing DT delays positive ROI for various companies [53, 55]. Apart from this cost of maintenance is very high.

1.6 ADVANTAGES OF DT

DTs can help manufacturing industries improve their strategies, models, and their operations [53, 57, 58]. In other sectors, there are several benefits that are listed below:

a. **Continuous Operations Monitoring:** DT can provide manufacturers with more deep insights and knowledge of their operation management. David Williamson who implemented DT in the IT industry said that DT can provide real-time tracking of performance and assets [55, 58].
b. **Enhanced Employee Performance:** In terms of better performance, the supervisor can check the performance more deeply and thoroughly in industries, for example, the performance of workers on an assembly line. This technology also enables supervisors to detect defects in the manufacturing process, for example, if the spinning time of an equipment is 5 minutes but it is rotating for 15 minutes. This issue can be checked by the supervisor with DT and can be corrected timely. In this way, the delay reason can also be easily detected which overall results in better performance of the product [].
c. **Enhanced Visions through Scenario Simulation:** This technology can predict the future performance of a particular product. After prediction, changes can be made at the time of planning. Let's suppose a user implements DT for a warehouse to keep track of robotic involvement with warehouse equipment [60, 64, 65]. Using DT, the user can check how to make the process fast and cost-effective.
d. **Cost Savings:** DT can also help in cost reduction due to its core involvement in manufacturing industries. It reduces company costs because there is no need to carry out real-time testing with physical prototypes. At the time of planning, we can ensure product requirements before final implementation that results in cost savings [34, 59, 66].

1.7 CONCLUSIONS

The Digital Twins (DT) has emerged as a transformational prototype that bridges the physical world to the virtual world in today's dynamic environment where technologies keep on changing. DT is a replica of a real world may be a system or process or object and simulation is the core component of DT which is used to implement the behavior of simulated object. This chapter provides an overview of simulation, DT and their types. This chapter also presents various market strategies such as the present situation, size, and share of DT in healthcare. Finally, this chapter concluded with challenging issues, benefits, and drawbacks of DT.

REFERENCES

1. Innovation Research. Available on: https://www.siemens.com/global/en/company/innovation/research-development/siemens-core-technologies/simulation-digital-twin.html [Accessed on 27 June 2024].
2. Market Research by KBV. Available on: https://www.kbvresearch.com/digital-twins-in-healthcare-market/ [Accessed on 20 August 2024].
3. Benefits and Drawback of DT. Available on: https://www.techtarget.com/searcherp/feature/Advantages-and-disadvantages-of-digital-twin-technology [Accessed on 12 August 2024].
4. Yoo SJ, Hussein N, Peel B, Coles J, Arsdell GS, Honjo O, Haller C, Lam CZ, Seed M, and Barron D. 3D modeling and printing in congenital heart surgery: entering the stage of maturation. *Frontiers in Pediatrics*, 9, 621672 (2021).
5. Anwar S, Singh GK, Miller J, Sharma M, Manning P, Billadello JJ, Eghtesady P, and Woodard PK. 3D printing is a transformative technology in congenital heart disease. *JACC: Basic to Translational Science*, 3(2), 294–312 (2018).
6. Sharma L, and Carpenter M (Eds.). *Computer Vision and Internet of Things: Technologies and Applications* (1st ed.). Chapman and Hall/CRC (2022). https://doi.org/10.1201/9781003244165
7. Bangare SL, Sharma L, Varade AN, Lokhande YM, Kuchangi IS, and Chaudhari NJ. Computer-Aided Lung Cancer Detection and Classification of CT Images Using Convolutional Neural Network. In *Computer Vision and Internet of Things: Technologies and Applications*, Taylor & Francis, CRC Press, pp. 247–262 (2022).
8. Singh J, Gupta D, and Sharma L. Analysis of Machine Learning Techniques for Airfare Prediction. In *Computer Vision and Internet of Things: Technologies and Applications,* Taylor & Francis, CRC Press, pp. 211–231 (2022).
9. Cooper J, Corrias A, Gavaghan D, and Noble D. Considerations for the use of cellular electrophysiology models within cardiac tissue simulations. *Progress in Biophysics and Molecular Biology*, 107, 74–80 (2011).
10. Carpenter M, Jangid VK, Carpenter D, and Sharma L. Innovation and Emerging Computer Vision and Artificial Intelligence Technologies in Coronavirus Control. In *Computer Vision and Internet of Things: Technologies and Applications*, Taylor & Francis, CRC Press, pp. 177–192 (2022).
11. Pathmanathan P, *et al.* A numerical guide to the solution of the bidomain equations of cardiac electrophysiology. *Progress in Biophysics and Molecular Biology*, 102, 136–155 (2010).
12. Hunter P, Pulian A, and Smail B. Modeling total heart function. *Annual Review of Biomedical Engineering*, 5, 147–177 (2003).

13. Sermesant M, *et al.* Patient-specific electromechanical models of the heart for the prediction of pacing acute effects in crt: A preliminary clinical validation. *Medical Image Analysis*, 16(1), 201–215 (2012).
14. Carpenter M, Carpenter D, Jangid VK, and Sharma L. Computer Vision in Surgical Operating Theatre and Medical Imaging. In *Computer Vision and Internet of Things: Technologies and Applications*, Taylor & Francis, CRC Press, pp. 75–96 (2022).
15. Sharma L, and Carpenter M. Use of Robotics in Real-Time Applications. In *Computer Vision and Internet of Things: Technologies and Applications*, Taylor & Francis, CRC Press, pp. 41–50 (2022).
16. Loppini A, *et al.* Competing mechanisms of stress-assisted diffusivity and stretch-activated currents in cardiac electromechanics. *Frontiers in Physiology*, 9, 1714 (2018).
17. Dedè L, Gerbi A, and Quarteroni A. Segregated algorithms for the numerical simulation of cardiac electromechanics in the left human ventricle. In *The Mathematics of Mechanobiology*, Springer, pp. 81–116 (2020).
18. Gillette K, *et al.* Automated framework for the inclusion of a his-purkinje system in cardiac digital twins of ventricular electrophysiology. *Annals of Biomedical Engineering*, 49, 3143–3153 (2021).
19. Sharma L. Human Detection and Tracking Using Background Subtraction in Visual Surveillance. In *Towards Smart World: Homes to Cities using Internet of Things*, Taylor & Francis, CRC Press, pp. 317–329 (2020).
20. Corti M, *et al.* Impact of atrial fibrillation on left atrium haemodynamics: A computational fluid dynamics study. *Computers in Biology and Medicine*, 150, 106143 (2022).
21. Garg PK. Smart E-healthcare with Internet of Things: Current Trends Challenges, Solutions and Technologies. In *From Visual Surveillance to Internet of Things*, Taylor & Francis, CRC Press2019).
22. Garg PK, and Agarwal N. A foresight on e-healthcare Trailblazers. In *From Visual Surveillance to Internet of Things*, Taylor & Francis, CRC Press (2019), pp. 210–211.
23. Viola F, Meschini V, and Verzicco R. Fluid-structure-electrophysiology interaction (fsei) in the left-heart: A multi-way coupled computational model. *European Journal of Mechanics - B/Fluids*, 79, 212–232 (2020).
24. Sharma L, and Garg PK. Future of Internet of Things. In *From Visual Surveillance to Internet of Things*, Taylor & Francis, CRC Press (2019).
25. Jha G, and Gupta S. E-health in Internet of Things (IoT) in Real-Time Scenario. In: Singh PK, Wierzchoń ST, Tanwar S, Ganzha M, Rodrigues JJPC (Eds.) *Proceedings of Second International Conference on Computing, Communications, and Cyber-Security*. Lecture Notes in Networks and Systems, Vol. 203. Springer, Singapore (2021). https://doi.org/10.1007/978-981-16-0733-2_48.
26. Sharma S, Verma S, and Kumar M. Use of Motion Capture in 3D Animation: Motion Capture Systems, Challenges, and Recent Trends. In *2019 International Conference on Machine Learning, Big Data, Cloud and Parallel Computing (COMITCon)*, 2019, pp. 289–294. https://doi.org/10.1109/COMITCon.2019.8862448.
27. Kumar S, Gupta P, and Lakra S. The Zeitgeist Juncture of "Big Data" and its Future Trends. In *2019 International Conference on Machine Learning, Big Data, Cloud and Parallel Computing (COMITCon)*, 2019, pp. 465–469. https://doi.org/10.1109/COMITCon.2019.8862433.
28. Makkar S, and Sharma L. A Face Detection Using Support Vector Machine: Challenging Issues, Recent Trend, Solutions and Proposed Framework. In: Singh M, Gupta P, Tyagi V, Flusser J, Ören T, Kashyap R (Eds.) *Advances in Computing and Data Sciences. ICACDS 2019. Communications in Computer and Information Science*, Vol. 1046. Springer, Singapore (2019). https://doi.org/10.1007/978-981-13-9942-8_1.

29. Sharma L, Sengupta S, and Kumar B. An Improved Technique for Enhancement of Satellite Images, *Journal of Physics: Conference Series*, 1714, 012051 (2021
30. Kumar A, Jha G, Sharma L, and Khatri SK. Challenges, Potential & future of IOT integrated with block chain. In *International Journal of Recent Technology and Engineering*, Vol. 8, Issue 2S7, pp. 530–536 (2019).
31. Anand A, Jha V, and Sharma L. An improved local binary patterns histograms techniques for face recognition for real time application. In *International Journal of Recent Technology and Engineering*, Vol. 8, Issue 2S7, pp. 524–529 (2019).
32. Sharma L, Yadav DK, and Singh A. Fisher's Linear Discriminant Ratio based Threshold for Moving Human Detection in Thermal Video. In *Infrared Physics and Technology*, Elsevier (2016) (SCI impact factor: 1.58, Published).
33. Sharma L, and Carpenter M (Eds.). *Computer Vision and Internet of Things: Technologies and Applications* (1st ed.). Chapman and Hall/CRC (2022).
34. Sharma L. *Towards Smart World: Homes to Cities using Internet of Things*, Taylor & Francis, CRC Press (2020).
35. Sharma L, and Garg PK. *From Visual Surveillance to Internet of Things*, Taylor & Francis, CRC Press (2019).
36. Sharma L, and Garg PK (Eds.). *Deep Learning in Internet of Things for Next Generation Healthcare* (1st ed.). Chapman and Hall/CRC (2024).
37. Sharma L, and Garg PK (Eds.). *Technological Prospects and Social Applications of Society 5.0* (1st ed.). Chapman and Hall/CRC (2023).
38. Corral-Acero J, Margara F, Marciniak M, Rodero C, Loncaric F, Feng Y, Gilbert A, Fernandes JF, Bukhari HA, Wajdan A, et al. The 'digital twin' to enable the vision of precision cardiology. *European Heart Journal*, 41, 4556–4564 (2020).
39. Stehlik J, Schmalfuss C, Bozkurt B, Nativi-Nicolau J, Wohlfahrt P, Wegerich S, Rose K, Ray R, Schofield R, Deswal A, et al. Continuous wearable monitoring analytics predict heart failure hospitalization. *Circulation: Heart Failure*, 13, e006513 (2020).
40. Sharma, L. (2020). The future of smart cities. In L. Sharma (Ed.), *Towards smart world: Homes to cities using Internet of Things* (pp. 1–19). CRC Press. https://doi.org/10.1201/9781003056751-1
41. Singh S, Sharma L, and Kumar B. A Machine Learning Based Predictive Model for Coronavirus Pandemic Scenario. *Journal of Physics: Conference Series*, 1714, 012023 (2021).
42. Saraogi G, Gupta D, Sharma L, and Rana A. Un-supervised approach to backorder prediction using deep Autoencoder. *Recent Advances in Computer Science and Communications (Formerly: Recent Patents on Computer Science)*, 14(8), 500–511 (2021).
43. Hussain I, Hossain MA, and Park S-J. A healthcare digital twin for diagnosis of stroke. In *Paper presented at: 2021 IEEE International Conference on biomedical engineering, computer and information technology for health (BECITHCON)*, IEEE, pp. 18–21 (2021).
44. Frangi AF, Rueckert D, Schnabel JA, and Niessen WJ. Automatic construction of multiple-object three-dimensional statistical shape models: application to cardiac modeling. *IEEE Transactions on Medical Imaging*, 21, 1151–1166 (2020).
45. Types of Digital Twins. Available at: https://program-ace.com/blog/types-of-digital-twins/ [Accessed on 20 September 2024].
46. Lloyd-Jones DM, Braun LT, Ndumele CE, Smith SC, Sperling LS, Virani SS, and Blumenthal RS. Use of risk assessment tools to guide decision-making in the primary prevention of atherosclerotic cardiovascular disease: a special report from the American Heart Association and American College of Cardiology. *Circulation*, 139, 139 (2019).

47. Digital Twin in Heart Surgeries. Available on: https://www.imperial.ac.uk/news/253154/digital-twin-heart-modelling-project-will/ [Accessed on 17 August 2024].
48. Wu H *et al.* A Comprehensive Review of Digital Twin from the Perspective of Total Process: Data, Models, Networks and Applications. *Sensors (Basel, Switzerland)*, 23(19), 8306 (2023). doi: 10.3390/s23198306.
49. Fuller A, Fan Z, Day C, and Barlow C. Digital Twin: Enabling Technologies, Challenges and Open Research. *IEEE Access*, 8, 108952–108971 (2020).
50. Lu Y, Peng T, and Xu X. Energy-efficient cyber-physical production network: Architecture and technologies. *Computers & Industrial Engineering*, 129, 56–66 (2019).
51. Shangguan D, Chen L, and Ding J. A Hierarchical Digital Twin Model Framework for Dynamic Cyber-Physical System Design. In *Proceedings of the 5th International Conference on Mechatronics and Robotics Engineering - ICMRE'19*, Rome, Italy, pp. 123–129, ACM Press (2019).
52. Tao F, Qi Q, Wang L, and Nee A. Digital Twins and Cyber–Physical Systems toward Smart Manufacturing and Industry 4.0: Correlation and Comparison. *Engineering*, 5, 653–661 (2019).
53. Tao F, Cheng J, Qi Q, Zhang M, Zhang H, and Sui F. Digital twindriven product design, manufacturing and service with big data. *The International Journal of Advanced Manufacturing Technology*, 94, 3563–3576 (2018).
54. Ruohomäki T, Airaksinen E, Huuska P, Kesäniemi O, Martikka M, and Suomisto J. Smart City Platform Enabling Digital Twin. In *2018 International Conference on Intelligent Systems (IS)*, pp. 155–161 (2018).
55. Liu J, Du X, Zhou H, Liu X, ei Li L, and Feng F. A digital twin-based approach for dynamic clamping and positioning of the flexible tooling system. *Procedia CIRP*, 80, 746–749 (2019).
56. Tao F, Zhang H, Liu A, and Nee AYC. Digital Twin in Industry: State-of-the-Art. *IEEE Transactions on Industrial Informatics*, 15, 2405–2415 (2019).
57. Huh S, Cho S, and Kim S. Managing IoT devices using blockchain platform. In *2017 19th International Conference on Advanced Communication Technology (ICACT)*, pp. 464–467 (2017).
58. Malik AA, Masood T, and Bilberg A. Virtual reality in manufacturing: immersive and collaborative artificial-reality in design of human-robot workspace. *International Journal of Computer Integrated Manufacturing*, 1, 22–37 (2020).
59. Negri E, Fumagalli L, and Macchi M. A Review of the Roles of Digital Twin in CPS-based Production Systems. *Procedia Manufacturing*, 11, 939–948 (2017).
60. Stark R, Fresemann C, and Lindow K. Development and operation of Digital Twins for technical systems and services. *CIRP Annals*, 68(1), 129–132 (2019).
61. Digital Twin in Heart Surgeries. Available on: https://engineering.jhu.edu/news/natalia-trayanovas-cardiac-digital-twins-featured-in-wall-street-journal/ [Accessed on 29 July 2024].
62. Novo O. Blockchain Meets IoT: An Architecture for Scalable Access Management in IoT. *IEEE Internet of Things Journal*, 5, 1184– 1195 (2018).
63. Soe RM. Smart Twin Cities via Urban Operating System. In *Proceedings of the 10th International Conference on Theory and Practice of Electronic Governance - ICEGOV '17*, New Delhi AA, India, pp. 391–400, ACM Press (2017).
64. Alam KM, and El Saddik A. C2ps: A Digital Twin Architecture Reference Model for the Cloud-Based Cyber-Physical Systems. *IEEE Access*, 5, 2050–2062 (2017).

65. Wu D, Jennings C, Terpenny J, and Kumara S. Cloud-based machine learning for predictive analytics: Tool wear prediction in milling. In *2016 IEEE International Conference on Big Data (Big Data)*, pp. 2062–2069 (2016).
66. Tao F, Cheng J, and Qi Q. IIHub: An Industrial Internet-of-Things Hub Toward Smart Manufacturing Based on Cyber-Physical System. *IEEE Transactions on Industrial Informatics*, 14, 2271–2280 (2018).

2 A Multilayered Architecture of Digital Twin Technology

Naba Suroor

2.1 INTRODUCTION

The Digital Twin (DT) concept is rapidly evolving, acting as a critical enabler in various technological fields, especially as industries move toward Industry 4.0. At its core, a Digital Twin serves as a dynamic virtual model of a physical system, object, or process, providing a real-time digital counterpart that reflects the characteristics and behaviors of its physical version. This innovation allows organizations to optimize operational efficiency, predict maintenance needs, and enhance the lifecycle management of products and systems [1]. From aerospace engineering to healthcare, manufacturing, and urban development, Digital Twins are redefining the future of data-driven decision-making. Additionally, the integration of advanced technologies such as Artificial Intelligence (AI), Machine Learning (ML), and the Internet of Things (IoT) into Digital Twins further amplifies their potential. These technologies enable Digital Twins to replicate physical entities and analyze vast amounts of data, predict outcomes, and simulate different scenarios in real time. This predictive capability allows for more accurate forecasting and decision-making, minimizing risks and optimizing processes across various sectors.

As Digital Twin technology continues to mature, its application expands beyond traditional industries, driving innovation in smart cities, autonomous vehicles, and personalized healthcare. A Digital Twin (DT) is essentially a digital replica of a physical entity, which could be a machine, a building, a human body, or even an entire production line. This virtual representation is continuously updated with dynamic data from sensors embedded in the physical system, allowing for precise monitoring, analysis, and simulation of its real-world counterpart. Some of the important key concepts of DT are listed below:

a. **Virtual Representation:** The DT is a computer-generated, highly detailed model of the physical object or system. It mirrors the physical twin in every aspect, from its geometry and design to its behavior under different operational conditions.

DOI: 10.1201/9781003582489-2

b. **Data-Driven:** The synchronization between the digital and physical twins is achieved through a continuous stream of data captured from sensors on the physical asset. These data streams can include a wide range of variables such as temperature, pressure, energy output, vibration, and environmental conditions [2].
c. **Simulation Capabilities:** A Digital Twin can simulate a wide array of operational scenarios and environmental conditions without disrupting the real-world asset. This allows engineers to test potential upgrades, anticipate failures, and make well-informed decisions to improve performance and extend the lifespan of the asset.
d. **Predictive Analytics:** Through the integration of Artificial Intelligence (AI) and Machine Learning (ML) algorithms, Digital Twins not only provide an actual view of an asset's current state but also offer predictive insights. These insights help predict when maintenance is required, what operational risks might arise, and how performance can be optimized over time.
e. **Bidirectional Communication:** DTs are not passive models. They maintain a feedback loop with the physical system. When a change is made to the physical system, the DT reflects this update, and vice versa. Actionable insights from the DT can influence modifications in the physical world [3].

2.2 PURPOSE AND FUNCTIONALITY OF DIGITAL TWINS (DTS)

The DT concept has revolutionized how organizations manage, monitor, and optimize physical assets. A DT serves as a high-fidelity, real-time digital replica of a physical system, combining data from sensors, simulations, and historical performance to create a dynamic representation of reality. This integration bridges the physical and digital worlds, enabling companies to monitor system conditions in factual time with unparalleled precision. Such visibility allows organizations to not only track the current state of their assets but also understand and predict how they will perform under different conditions. As a result, the DT acts as a decision-making powerhouse, offering actionable insights that can streamline operations and enhance overall productivity.

One of the most transformative aspects of DT technology is its ability to simulate and optimize performance in a virtual environment. Using historical and real-time data, organizations can test various operational scenarios without disrupting real-world systems. This capability enables companies to evaluate the impact of potential changes, such as adjusting system configurations or introducing new processes, all in a risk-free manner. By modeling different conditions and analyzing outcomes, businesses can fine-tune their strategies, improve resource utilization, and enhance efficiency. Furthermore, DTs can identify inefficiencies or malfunctions that might escape human detection. By analyzing subtle deviations in performance, they ensure that small issues are resolved before escalating into costly problems, helping organizations maintain seamless operations.

DTs go beyond optimization and monitoring by introducing a predictive and preventive dimension to asset management. Predictive maintenance, powered by DTs, allows companies to anticipate equipment failures or maintenance needs by analyzing trends and patterns in the data. This proactive approach minimizes unplanned downtime, reduces maintenance costs, and extends the lifespan of critical assets. Additionally, DTs provide valuable insights into asset behavior over time, helping organizations refine designs, improve safety, and ensure reliability. These capabilities represent a significant departure from traditional methods of asset management, which often relied on reactive measures or periodic inspections. Instead, DTs enable organizations to adopt a forward-looking, data-centric approach that not only improves performance but also drives innovation and resilience in a competitive landscape [4].

2.3 THE HISTORY AND EVOLUTION OF DIGITAL TWIN

The origins of the DT concept are deeply rooted in simulation and modeling, but its current application and relevance were spurred by advancements in data processing, Internet of Things (IoT) technologies, and machine learning. Below is a detailed timeline of the development of DTs.

2.3.1 1960s: Early Predecessors in Aerospace

While the formal concept of a DT did not emerge until the 21st century, the core idea has existed for decades, especially in aerospace engineering. During the Apollo space missions of the 1960s, NASA engineers used physical replicas of spacecraft, known as analogs, to monitor and predict the behavior of spacecraft systems in orbit. For instance, ground teams used earth-bound copies of spacecraft modules to simulate and troubleshoot operational issues faced by astronauts.

This practice can be seen as the earliest form of digital-physical mirroring that characterizes modern DT technology, although at that time, the technology was not fully digital.

2.3.2 2002: The Conceptual Birth of DTs

The formal concept of a DT was introduced in 2002 by Dr. Michael Grieves, who was a faculty member at the University of Michigan at the time. Dr. Grieves conceptualized the idea within the context of Product Lifecycle Management (PLM), describing the use of digital models to mirror the physical-world lifecycle of products and systems. This digital replica would act as a virtual counterpart to the physical asset, enabling engineers to predict performance, simulate design changes, and optimize operations. Dr. Grieves' vision emphasized that digital replicas could be used to simulate, analyze, and optimize the lifecycle of products, from design and manufacturing to deployment and maintenance. The DT could be used across the entire product lifecycle, helping companies improve product quality and operational efficiency [5].

2.3.3 2010: NASA's Adoption and Terminology

In 2010, John Vickers from the National Aeronautics and Space Administration (NASA), USA, further developed the concept by coining the term DT. NASA adopted DT technology to monitor the health and performance of its space vehicles and systems in orbit, where maintenance is not easily accessible. The DT concept allowed NASA to simulate mission-critical systems in real time and predict potential failures before they occurred.

NASA's adoption of the DT paradigm marked a significant step forward, demonstrating the practical value of virtual models in practical operations.

2.3.4 2010s–Present: Expansion across Industries

The rise of Internet of Things (IoT) devices, big data, cloud computing, and advanced machine learning techniques in the early 2010s catalyzed the widespread adoption of DT technology. As IoT sensors became more affordable and cloud infrastructure more accessible, companies across various sectors began using DTs to optimize their operations. By 2020, DTs had grown into a $5 billion market, with widespread applications across manufacturing, healthcare, automotive, smart cities, and energy sectors [6]. The technology was no longer limited to high-stakes fields like aerospace and defense; instead, it became a key part of Industry 4.0, enabling data-driven decision-making and continuous improvement in a variety of environments.

2.4 DT ARCHITECTURE: A MULTI-LAYERED SYSTEM

A DT's architecture is designed to enable current data flow, seamless simulation, and decision-making capabilities. It involves several interconnected layers and components, all of which work together to create a reliable and responsive digital counterpart of the physical system.

Figure 2.1 provides a layered representation of the DT system, demonstrating how physical assets integrate with digital models for real-time monitoring and decision-making. It begins with the Physical Entity, representing the real-world object or system, followed by the Data Collection Layer (IoT and sensors), which gathers relevant information. This data is transmitted through the Communication Layer (network and cloud), forming the backbone of connectivity.

At the core is the Digital Twin Model (modeling and simulation), where real-world data is analyzed and mirrored virtually. This feeds into the Processing Layer (data analytics and AI), which extracts actionable insights. Finally, the User Interface & Visualization layer displays these insights for operators or decision-makers, forming a Feedback Loop that ensures continuous improvement and optimization of the physical system.

2.4.1 Core Components of DT Architecture

a. **Physical Entity:** This is the real-world object or system that is being mirrored in the digital space. It could be a machine, an entire factory, a vehicle,

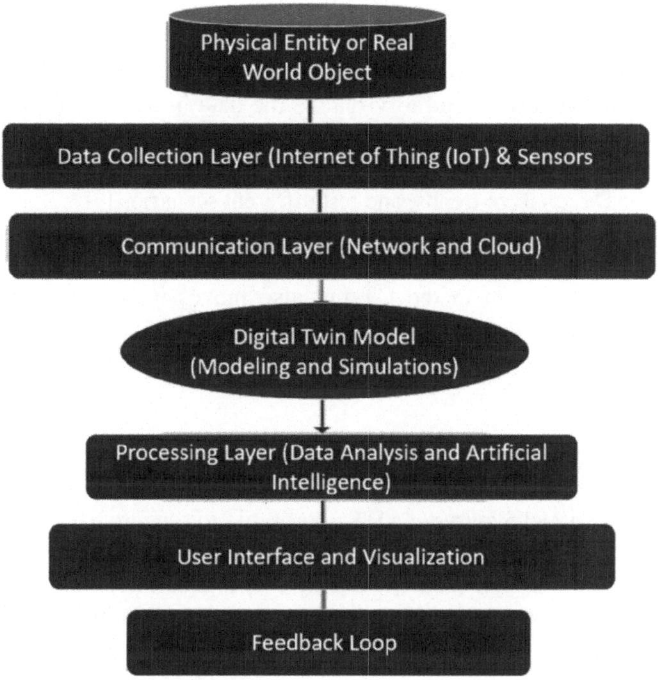

FIGURE 2.1 Layered architecture of a Digital Twin system

or even a human body. The physical entity is embedded with sensors to capture data on its performance and condition.

b. **Data Collection Layer:** Sensors and IoT devices are placed on the physical object to continuously gather data related to its operation. This could include temperature, humidity, pressure, energy output, or environmental conditions [7]. This layer also ensures that data is transmitted securely to the cloud or processing system.
c. **Communication Layer:** The data captured by the sensors is sent to the cloud via secure networks. This layer handles the transfer of simultaneous data between the physical system and the DT, ensuring that information flows both ways – data from the physical system is fed into the DT, and insights from the twin can be relayed back to the physical system.
d. **Modeling and Simulation Layer:** The core of the DT resides here – a highly detailed virtual model of the physical system is built using computer-aided design (CAD), finite element analysis (FEA), and other digital modeling techniques [8]. This layer allows for real-time and predictive simulations of the physical object's behavior under various conditions.
e. **Data Processing and Analytics Layer:** At this level, data collected from the physical object is analyzed using advanced AI and machine learning algorithms. This layer is responsible for generating insights, identifying potential inefficiencies or failures, and making predictions about future

behavior. The data processing and analytics layer is where much of the "intelligence" of a DT is realized.

f. **User Interface and Visualization Layer:** To make the insights from the DT actionable, users need to visualize the data in intuitive formats. This layer consists of dashboards, 3D models, and augmented or virtual reality (AR/VR) interfaces that allow operators and decision-makers to interact with the DT, analyze trends, and test different scenarios [9].
g. **Feedback Loop:** A crucial aspect of the DT architecture is the bi-directional feedback loop. This means that data not only flows from the physical system to the DT but also moves in the opposite direction. When actionable insights are derived from the DT, they can be fed back to the physical system, enabling automated or manual adjustments dynamically.

2.5 REAL-TIME DATA INTEGRATION AND SYNCHRONIZATION

One of the most critical features of a DT is its ability to synchronize with its physical counterpart in real time. This continuous data flow enables a high degree of accuracy and precision in the digital model, ensuring that it accurately reflects the state of the physical object at any given moment. By constantly updating the digital model with real-world data, organizations can monitor, simulate, and optimize operations as they unfold [10].

2.5.1 AI and Machine Learning Integration

The true predictive and diagnostic power of a DT lies in its seamless integration with artificial intelligence (AI) and machine learning (ML). By leveraging these technologies, DTs go beyond simple data representation to deliver actionable insights that can transform operations. Machine learning models are trained on extensive datasets comprising historical performance records, real-time sensor data, and even environmental variables. This training enables the DT to identify patterns, predict future outcomes, and detect anomalies that might escape human attention or traditional monitoring tools. Such capabilities are instrumental in enhancing operational efficiency, reducing downtime, and improving overall system reliability [11].

In the context of a manufacturing plant, for example, a DT equipped with AI and ML can analyze historical data on equipment performance to identify subtle signs of wear or potential failure. It can forecast when a machine is likely to fail, providing ample warning to schedule preventive maintenance. This proactive approach not only avoids unexpected downtime but also minimizes costly emergency repairs. Furthermore, AI-powered DTs can simulate various operational scenarios, enabling engineers to test strategies for optimization in a virtual environment before implementing them in the real world. This ensures better decision-making, cost efficiency, and resource allocation.

What sets AI-driven DTs apart is their ability to learn and evolve. Unlike static systems, AI continuously refines its algorithms based on new data inputs, making the

DT more intelligent and adaptable to complex environments. This evolving intelligence is especially valuable in dynamic industries such as aerospace, healthcare, and energy, where systems are subject to constantly changing conditions. For instance, in the energy sector, AI-integrated DTs can optimize power distribution, predict outages, and adapt to fluctuating demand patterns. Overall, the incorporation of AI and ML transforms DTs into powerful tools that not only replicate physical systems but actively enhance their performance and longevity.

2.6 TYPES OF DIGITAL TWIN

The scope of a DT can vary greatly depending on the complexity of the physical object or system it is designed to represent. DTs are typically categorized into different types based on their level of magnification or application. The main components of DT are given below.

2.6.1 Component Twins (or Parts Twins)

At the most granular level, component twins represent individual components of a larger system. For example, in a jet engine, the DT of a turbine blade would be considered a component twin [12]. These twins are often used to monitor the performance of critical components, detect micro-level issues, and optimize the functioning of individual parts. By focusing on specific components, issues at the earliest stages can be identified, before they affect the overall system.

2.6.2 Asset Twins

Asset twins represent the next level of magnification and model the interaction of multiple components that work together as a complete asset. For example, the jet engine itself is an asset twin made up of several component twins, such as the turbines, compressors, and combustion chambers. Asset twins allow companies to monitor how these components interact as a cohesive system and provide insights into the overall health and performance of the asset [13]. By modeling the interaction between components, asset twins help in optimizing system-wide performance and identifying inefficiencies or bottlenecks that may arise from the interaction of different parts.

2.6.3 System or Unit Twins

System twins are the next level in the hierarchy and are designed to model the interaction of multiple assets within a broader system. For example, in a power plant, a system twin would simulate the interaction between various assets like turbines, generators, and cooling systems. This allows for a more comprehensive understanding of how different assets work together, providing insights into the overall operation of the entire system. System twins are often used to model large-scale systems, such as industrial manufacturing processes or infrastructure networks. These twins provide a macro-level view of how individual assets interact with each other and how changes to one asset might affect the performance of the entire system.

2.6.4 Process Twins

At the highest level of magnification, process twins represent an entire production line or industrial process. A process twin enables companies to simulate, monitor, and optimize complex workflows that involve multiple systems and assets. For example, in a manufacturing plant, a process twin might model the entire assembly line, allowing operators to optimize the timing of different processes to minimize delays and maximize efficiency. Process twins are especially useful in industries with complex supply chains or production processes that involve multiple steps or systems. By understanding how these processes interact, companies can streamline operations, reduce inefficiencies, and make more informed decisions about how to improve their workflows.

2.7 DIGITAL TWINS VS. TRADITIONAL SIMULATIONS

Although simulations and DTs share similarities in their use of digital models, they differ significantly in scope, functionality, and capabilities. The most critical difference lies in the integration of real-time data and the bidirectional nature of DTs.

2.7.1 Traditional Simulations

A simulation is a one-time, often static, study of a particular process or system. Simulations rely on predefined parameters and variables, and once initiated, they provide results based on these fixed inputs. Traditional simulations are typically used to study specific scenarios or optimize certain operations [14]. Taking the instance of the design phase of an aircraft engine, simulation may be used to understand how the engine will perform under certain conditions, such as high altitudes or extreme temperatures. However, traditional simulations do not update immediately and do not incorporate live data from the actual engine in operation.

2.7.2 Digital Twins: A Continuous, Real-Time Feedback Loop

A DT, on the other hand, goes beyond simulations by maintaining a two-way flow of information between the physical object and its virtual counterpart. This allows the DT to continuously update its model with real-time data, creating a highly dynamic and accurate representation of the physical object. Because DTs are constantly fed with live data from sensors, they can study a wide range of scenarios instantaneously [15]. This provides far more insights and capabilities than a one-time simulation. For example, in a wind turbine, the DT can simulate how changing weather conditions will affect energy output, optimize the turbine's angle for maximum efficiency, and predict when a component is likely to fail [16]. DTs can run multiple simulations concurrently, studying how various conditions impact different parts of the system, which enables much broader and more detailed analyses than traditional simulations.

2.8 CHALLENGES IN IMPLEMENTING DIGITAL TWIN TECHNOLOGY

While the concept of DTs is compelling and beneficial, its full-scale implementation is not without significant hurdles [17]. The primary challenges revolve around data management, computational power, integration, cybersecurity, and scalability.

2.8.1 Data Integration Complexity

One of the most critical challenges for deploying a DT is the seamless integration of data from a variety of sources. DTs rely on data collected from sensors, IoT devices, and other digital systems that feed into a central model. However, many industries have complex ecosystems, with multiple legacy systems and devices from different vendors that may not easily communicate with each other.

a. **Heterogeneous Systems:** Integrating data from disparate systems into a unified DT architecture can be a significant technical challenge.
b. **Data Standardization:** There is often a lack of standardization in how data is collected and processed, leading to inconsistencies that make it difficult to build an accurate digital model.
c. **Real-Time Synchronization:** Achieving simultaneous synchronization between the physical entity and its DT is essential, but it requires advanced algorithms and high-speed communication networks. Delays or inaccuracies in data synchronization can compromise the effectiveness of a DT and lead to incorrect predictions or decisions.

2.8.2 Scalability Issues

Maintaining its efficiency and accuracy becomes increasingly difficult as more components, processes, or systems are added to a DT model [18]. The scalability challenge is particularly evident when moving from small, localized DT implementations (such as for a single machine) to larger-scale systems (such as an entire factory, a transportation network, or even a smart city).

a. **Computational Load:** Simulating large-scale systems at present requires immense computational resources [19].
b. **Data Overload:** As the number of devices and sensors increases, the volume of data grows exponentially. Handling this data deluge can overwhelm traditional data processing systems, leading to performance bottlenecks [20].
c. **Managing Complexity:** As DTs scale to represent more complex systems, managing the interdependencies between different components and systems becomes a significant challenge. Even a small error in one part of the system can have ripple effects that disrupt the entire DT model.

2.8.3 Computational Power Requirements

Running a DT involves the continuous processing of vast amounts of data, and the execution of simulations instantaneously. This requires significant computational power:

a. **High-Fidelity Simulations:** For highly detailed and complex models, such as those used in aerospace engineering or automotive crash testing, DTs require high-fidelity simulations. These simulations must account for thousands of variables and operate under a wide range of conditions, all of which consume significant computational resources [21].
b. **Energy Consumption:** The energy cost of running DTs, especially at scale, is another challenge. Large cloud data centers running continuous simulations for multiple DTs consume a considerable amount of power, which raises questions about sustainability and cost efficiency [22, 23].

2.8.4 Interoperability and Standardization

One of the significant obstacles to the widespread adoption of DTs is the lack of global standards and interoperability frameworks. Different industries and vendors use their own formats, protocols, and technologies, which complicates the process of integrating various systems into a unified DT [24, 25, 26].

a. **Industry-Specific Standards:** Without universal standards, it is difficult to create scalable DT solutions that work across multiple industries or geographies [27, 28].
b. **Vendor Lock-In:** Many DT solutions are proprietary, meaning that organizations are often locked into using a single vendor's software or hardware ecosystem. This limits the flexibility to integrate new systems or technologies in the future [29, 30].

2.9 THE FUTURE OF DIGITAL TWIN

As industries continue to embrace DT technology, the global DT market is poised for exponential growth. The future of DTs holds immense promise, with new advancements likely to push the boundaries of what this technology can achieve [31].

a. **Autonomous Systems and Self-Healing Technologies:** As AI becomes more integrated with DTs, the technology will increasingly be used to develop autonomous systems capable of operating independently [32–33, 34, 35]. In the future, self-healing systems that can detect and repair faults without human intervention could become the norm.
b. **Expansion Beyond Industry 4.0:** While DTs are currently most closely associated with Industry 4.0, their future applications will expand far

beyond manufacturing. Sectors such as healthcare, education, urban planning, and climate science are already beginning to adopt DTs, and this trend is expected to continue [36, 37].

c. **Quantum Computing:** As quantum computing matures, it will likely play a transformative role in the future of DTs. The massive processing power of quantum computers could allow DTs to simulate incredibly complex systems at an unprecedented scale and speed [24].

2.10 CONCLUSION

The DT represents a groundbreaking evolution in how industries manage, monitor, and optimize physical systems. By bridging the physical and digital worlds, DTs create high-fidelity virtual replicas of real-life assets, allowing organizations to gain actionable insights and achieve unprecedented levels of operational excellence. These virtual models provide the ability to monitor physical systems in real time, simulate scenarios, predict failures, and optimize performance, making DTs an indispensable tool in today's data-driven environment. The combination of Internet of Things (IoT) devices, artificial intelligence (AI), machine learning (ML), and advanced analytics has positioned DTs at the forefront of technological innovation, enabling industries to adapt to challenges dynamically and harness the full potential of their resources.

The advantages of DTs are vast and impactful. First and foremost, they enhance operational efficiency by providing a real-time view of system performance, enabling rapid decision-making. Predictive capabilities help in identifying potential failures or inefficiencies, reducing downtime, and extending asset lifespans. Furthermore, DTs allow for risk-free experimentation through simulation, enabling organizations to explore various operational strategies without real-world disruptions. Their ability to integrate data from multiple sources makes them a powerful tool for addressing complex challenges in asset management, logistics, and supply chain optimization. Additionally, the growing integration of AI and ML ensures that DTs continuously learn and improve, becoming more intelligent and capable over time.

However, DTs are not without their challenges. One significant disadvantage is the complexity and cost associated with their implementation. Building and maintaining a DT requires substantial investments in hardware, software, and expertise. Additionally, the reliance on IoT sensors and networks introduces vulnerabilities, such as cybersecurity risks and potential data breaches. The accuracy of a DT is also dependent on the quality of the data it receives, making robust data governance and validation essential.

In summary, DTs signify a transformative shift in industry operations by introducing a predictive, dynamic, and data-centric approach to managing physical systems. While the advantages far outweigh the challenges, careful planning, and robust infrastructure are required to maximize their potential. As advancements in IoT, AI, and computing technologies continue to accelerate, DTs will evolve further, shaping the future of industries and unlocking new horizons of innovation.

REFERENCES

1. IBM. "What is a digital twin?" Available at: https://www.ibm.com/topics/what-is-a-digital-twin [accessed on October 5, 2024].
2. Microsoft Azure. "Azure Digital Twins: An IoT Platform for Real-Time Data." Available at: https://azure.microsoft.com/en-us/services/digital-twins/ [accessed on October 5, 2024].
3. International Organization for Standardization (ISO). "Smart Manufacturing and Digital Twin Standardization." Available at: https://www.iso.org/committee/7203984.html [accessed on October 5, 2024].
4. IEEE Xplore. "Digital Twin in Industry 4.0: Technologies, Applications and Challenges." Available at: https://ieeexplore.ieee.org/document/8972134 [accessed on October 5, 2024].
5. Bosch Rexroth. "Digital Twin Technology in Industry 4.0." Available at: https://www.boschrexroth.com/en/xc/industries/factory-automation/digital-twin/ [accessed on October 5, 2024].
6. Gartner. "Definition of Digital Twin." Available at: https://www.gartner.com/en/information-technology/glossary/digital-twin [accessed on October 5, 2024].
7. ScienceDirect. "Review of digital twin about concepts, technologies, and industrial applications." Available at: https://www.sciencedirect.com/science/article/pii/S0278612520301072 [accessed on October 5, 2024].
8. MIT Sloan Management Review. "The Digital Twin Opportunity." Available at: https://sloanreview.mit.edu/article/the-digital-twin-opportunity/ [accessed on October 5, 2024].
9. Milton, Castulo, et al. "Controller-Embeddable Probabilistic Real-Time Digital Twins for Power Electronic Converter Diagnostics." Available at: https://www.researchgate.net/publication/339061583_Controller-Embeddable_Probabilistic_Real-Time_Digital_Twins_for_Power_Electronic_Converter_Diagnostics [accessed on October 5, 2024].
10. ScienceDirect. "Integration of real-time locating systems into digital twins." Available at: https://www.sciencedirect.com/science/article/pii/S2452414X20300492 [accessed on October 5, 2024].
11. Siemens. "Digital Twin." Available at: https://www.sw.siemens.com/en-US/technology/digital-twin/ [accessed on October 5, 2024].
12. Deloitte. "Industry 4.0 and the digital twin." Available at: https://www2.deloitte.com/content/dam/Deloitte/kr/Documents/insights/deloitte-newsletter/2017/26_201706/kr_insights_deloitte-newsletter-26_report_02_en.pdf [accessed on October 5, 2024].
13. General Electric (GE). "Digital Twin Software." Available at: https://www.ge.com/digital/applications/digital-twin [accessed on October 5, 2024].
14. OECD. "Sustainability and Digital Twin Technology: Monitoring Environmental Impact." Available at: https://www.oecd.org/innovation/digital-twin-sustainability/ [accessed on October 5, 2024].
15. Accenture. "How digital twins enable autonomous operations." Available at: https://www.accenture.com/us-en/insights/industry-x/manufacturing-systems-architecture [accessed on October 5, 2024].
16. World Economic Forum. "Digital twins: What are they and why do they matter?" Available at: https://www.weforum.org/agenda/2022/05/digital-twin-technology-virtual-model-tech-for-good/ [accessed on October 5, 2024].
17. RTIInsights. "Another Avenue for Digital Twins: Behavioral Modeling for Banks." Available at: https://www.rtinsights.com/another-avenue-for-digital-twins-behavioral-modeling-for-banks/ [accessed on October 5, 2024].

18. SpringerLink. “Empirical Insights into the Challenges of Implementing Digital Twins.” Available at: https://link.springer.com/chapter/10.1007/978-981-19-3455-1_18 [accessed on October 5, 2024].
19. PwC. “How digital twins can make smart cities better.” Available at: https://www.pwc.com/m1/en/publications/how-digital-twins-can-make-smart-cities-better.html [accessed on October 5, 2024].
20. Harvard Business Review. “Digital Twins Can Help You Make Better Strategic Decisions.” Available at: https://hbr.org/2024/09/digital-twins-can-help-you-make-better-strategic-decisions [accessed on October 5, 2024].
21. Yu, W., et al. (2022). “Energy digital twin technology for industrial energy management: Classification, challenges and future.” *Renewable and Sustainable Energy Reviews*, 161, 112407. https://doi.org/10.1016/j.rser.2022.112407.
22. Javaid, M., et al. (2023). “Digital twin applications toward Industry 4.0: A review.” *Cognitive Robotics*, 3, 71–92. https://doi.org/10.1016/j.cogr.2023.04.003.
23. Semeraro, C., et al. (2023). “Digital twin application in energy storage: Trends and challenges.” *Journal of Energy Storage*, 58, 106347. https://doi.org/10.1016/j.est.2022.106347.
24. Rasheed, A., et al. (2019). “Digital twin: Values, challenges, and enablers from a modeling perspective.” *IEEE Access*, 8, 21980–22012. 10.1109/ACCESS.2020.2970143.
25. Lavanya Sharma et.al., “An Overview of Security Issues of Internet of Things”, *Computer Vision and Internet of Things: Technologies and Applications”*, Taylor & Francis, CRC Press, pp.29-40, May 2022
26. Lavanya sharma, “Rise of Computer Vision and Internet of Things”, *Computer Vision and Internet of Things: Technologies and Applications”*, Taylor & Francis, CRC Press, pp.5-17, May 2022.
27. Khan, L. U., et al. (2022). “Digital twin of wireless systems: Overview, taxonomy, challenges, and opportunities.” *IEEE Communications Surveys & Tutorials*, 24(4), 2230–2254. https://doi.org/10.1109/COMST.2022.3198273.
28. Sharma, L., & Garg, P.K. (Eds.). (2023). Technological Prospects and Social Applications of Society 5.0 (1st ed.). Chapman and Hall/CRC. https://doi.org/10.1201/9781003324720
29. Liu, M., et al. (2021). “Review of digital twin about concepts, technologies, and industrial applications.” *Journal of Manufacturing Systems*, 58, 346–361. https://doi.org/10.1016/j.jmsy.2020.06.017.
30. Sharma, L., & Carpenter, M. (Eds.). (2022). Computer Vision and Internet of Things: Technologies and Applications (1st ed.). Chapman and Hall/CRC. https://doi.org/10.1201/9781003244165
31. Sharma, L., & Garg, P.K. (Eds.). (2021). Artificial Intelligence: Technologies, Applications, and Challenges (1st ed.). Chapman and Hall/CRC. https://doi.org/10.1201/9781003140351
32. Barricelli, B. R., et al. (2019). “A survey on digital twin: Definitions, characteristics, applications, and design implications.” *IEEE Access*, 7, pp.167653-167671. https://doi.org/10.1109/ACCESS.2019.2953499.
33. Fuller, A., et al. (2020). “Digital twin: Enabling technologies, challenges, and open research.” *IEEE Access*, 8, 108952–108971. https://doi.org/10.1109/ACCESS.2020.2998358.
34. Sharma, L. (Ed.). (2020). Towards Smart World: Homes to Cities Using Internet of Things (1st ed.). Chapman and Hall/CRC. https://doi.org/10.1201/9781003056751
35. Sharma, L., & Garg, P.K. (Eds.). (2019). *From Visual Surveillance to Internet of Things: Technology and Applications* (1st ed.). Chapman and Hall/CRC. https://doi.org/10.1201/9780429297922 .

36. Shahzad, M., et al. (2022). “Digital twins in built environments: An investigation of the characteristics, applications, and challenges.” *Buildings*, 12(2), 120. https://doi.org/10.3390/buildings12020120.
37. Wang, Z., et al. (2022). “Mobility digital twin: Concept, architecture, case study, and future challenges.” *IEEE Internet of Things Journal*, 9(18), 17452–17467. https://doi.org/10.1109/JIOT.2022.3156028.

3 Understanding Data Framework for Digital Twinning

Rajeev Kumar, Balaji Aryan Singh, and Mrityunjay Sharma

3.1 INTRODUCTION

Digital twins are virtual models of a physical asset, system, or process and are used to simulate, predict, and optimize performance. Thus, a number of data types will be required to assure the accuracy and reliability of the Digital Twin. This recognition of what is needed in terms of defining these data requirements is vital for successfully implementing Digital Twins since they provide a foundation for building actual and useful models. Digital twin technology is finding powerful use cases across many industries. For the manufacturing and AEC sectors, digital twins of buildings provide teams with a deeper understanding of their physical space across its lifecycle-from initial designs to construction progress, then ongoing use and maintenance [1].

In this multicultural field, we will understand this topic with the help of some startups who have been working in the same. One of them is a startup called **Augtual Reality Labs Pvt Ltd** incubated at Atal Incubation Centre-BIMTECH, essentially working on a digital twin concept, Fabrik, which is a digital twin metaverse platform for industrial transformation for critical and high-value assets in aerospace, defense, oil and gas, automotive, and power sectors. Fabrik saves a lot of time and also reduces losses for high-capex equipment and processes by placing a digital twin over their physical and digital infrastructures. There is a need to operationalize multiple types of data. Proper understanding and definition of data requirements can lead to the creation of realistic, reliable, and useful Digital Twins. Strong data management, including collection, storage, processing, and analysis of data, can tap into the actual value of Digital Twins [1, 2].

Startups now increasingly play a big role in supporting and enabling Digital Twins in various domains. For instance, a startup called **Edgeforce Solutions Pvt Ltd** is developing a product in defense and security. Edgeforce was formed with the aim of introducing advanced and disruptive technology for the security forces in the country. As threats from inside and outside the country grow, it is imperative to develop technological advantages that can help us win conflicts in future. We can

DOI: 10.1201/9781003582489-3

achieve this by creating "in-house capabilities" that are innovative and futuristic, especially with regard to applications of Edge Technology. The output of such an initiative would then be the creation of solutions that could be impactful enough to support other government sectors such as PSUs, Logistics, Ports, and Mines. We're proud to be **Industry 4.0** [3], capable of building modern connected Digital Twins for physical assets. *Many start-ups keep busy in enhancing and developing new technologies for Digital Twin, concentrating on the integration and the management of data requirements.*

3.2. TYPES OF DATA

In order to effectively implement Digital Twins, several types of data are essential. These data types ensure the accuracy, reliability, and functionality of the digital representations of physical assets, systems, or processes. Here are the key data types needed for Digital Twins.

3.2.1 Historical Data

Now that we have learned our basic knowledge of what the requirements of data for Digital Twins are, let's jump right into one of the most important forms of the requirements of data. There's such great rage and flashing developments in technology that may leave us missing one crucial element behind all these developments. At the core of these developments lie Historical Data. That's correct. Even if one sees this as only "old data," it's, in fact, so much more. It could almost be considered a backbone of the entire Digital Twin notion. It's really the underpinning for effective modeling of a Digital Twin: a comprehensive view of an asset's life cycle performance. And this is just what Historical Data offers up: the whole saga-good, bad, and all the ugly stuff. So there are maintenance records, performance statistics, and usage logs – and that is only the beginning. All that data isn't simply where it has been, but what all of this portends. Well known is the fact that the better decision requires knowing something about the past of the asset. For example, knowing when machines broke down in the past helps avoid such incidents from occurring again in the future in a factory. This is where Historical Data proves very useful. Patterns emerge that show crucial details, like finding out that a machine consistently breaks down after 100 hours of operation. With historical data, the decision-making approach changes from guessing to using verifiable data that may save much time and resource.

Historical Data also makes the last benchmark for prediction. It will become next to impossible for the predictive models if they cannot analyze previous trends to foresee outcomes. Once there are robust historical data, it starts making sense about predictions. A proactive status will also empower one in being able to predict where the potential problem may appear before it manifests into the costly disturbance phase. Although historical data would appear like just numbers, it is actually the complete history behind the numbers. It can be referred to as the asset life story full of ups and downs, successes, and failures. It is even more pertinent for long lifecycle assets like infrastructure or aerospace. It is invaluable to record an asset's performance to take proper decisions at a later stage. To put it bluntly, Historical Data is the unsung hero of Digital Twins. They enable an understanding of context associated with an asset, informing better decisions today and making accurate future projections possible, but most importantly, it is going to become ever so vital as the Digital Twin technology advances and will evolve in its own right. It serves as the base that makes these digital models just as accurate and reliable as they should be.

Gemini Technology is a startup company providing a cloud-based platform that enables businesses to create, manage, and analyze the Digital Twins of their assets. The company focuses on accuracy and reliability in data that is constantly upgraded in validation and processing with the help of advanced tools for the enhancement of integration services. Solutions from Gemini Technology seem fairly promising for the gamut of industrial applications ranging from energy to transportation and subsequently to infrastructure management. Therefore, the company is set to gain importance in various sectors.

3.2.2 Real-Time Data

Real-Time Data captures the "now," that is, it can record a machine on the shop floor of a factory, the whole production line, or even the climate controls of a smart building. Of course, detection should occur as it happens instead of discovering it later on. This is why Real-Time Data is essential – it is the heartbeat of the digital twin, continually fed with real-time feeds of events as they happen in real time. Real-Time Data continually feeds the system with the live state of physical assets including sensor readings, operational metrics, and environmental conditions. Since Historical Data is the tale of what has already been done, Real-Time Data would be like a real-time television broadcast of current situations. Such sensors could include temperature, pressure, or vibration sensors with continuous real-time updates, 24/7. Changes often happen fast – a machine can overheat, or vibrations may rise above normal on some equipment. Real-Time Data enables immediate detection of such anomalies, allowing for quick responses without waiting for damage to accrue. It is used not only to catch problems but also to ensure the highest performance of the operation through the monitoring of metrics and making adjustments in real time. From the speed of a conveyor belt to cooling balance in a data center, operators are allowed to have real-time data for optimal performance. The data-driven control allows the operations to run smoothly without interruption. On a broader level, Real-Time Data facilitates continuous optimization of performance. Regular monitoring of assets offers operational fine-tuning data-continuous waste-eating operations that boost productivity. Ongoing data supply provides any organization with full visibility about its operations to enable on-time decisions based on concrete events [3–5].

In a nutshell, Real-Time Data brings to life the Digital Twin, bridging the gap between the digital model and the real-world scenario. It eliminates discrepancies in what is shown on the digital model and what really is in the real world. Without Real-Time Data, the system would be blind, only responding to problems once they are worse. With it, there is always an ongoing sense of conditions that can be addressed at the earliest sign of an issue. For instance, **Cognition Factory** is a relatively young startup that focuses on the processing and management of real-time data within Digital Twins. The company produces highly responsive and accurate Digital Twins with special focus on manufacturing, logistics, and smart city applications through the co-integration of IoT sensor data, legacy systems, and others. Optimization of factory automation is offered by Cognition Factory, using robotics, machine learning, and predictive analytics in order to achieve efficiency of material flow with the use of AMRs and AGVs within the smart factory and warehouse. One such success story incubated at the **Atal Incubation Centre** is **FlyatMach**, which relies on real-time data. FlyatMach is a precision agriculture company that helps farmers in crop management with insights derived from satellites, GPS, or drones. Disaster UAV expertise from FlyatMach has enabled quick assessment and sustainable solutions to happen during natural and man-made disasters for search and rescue activities. The evolving and responsive Digital Twin, therefore, makes real-time data pivotal to applications ranging from disaster management to smart factories that keep pace with fluctuating conditions of the real world.

3.2.3 Static Data

An example of static data is **Dataspire Technologies Private Limited**, a startup incubated at **CODISSIA Defence Innovation** and **Atal Incubation Centre**. Dataspire is an IT consultancy and outsourcing firm, concentrating on Database and Infrastructure Management. It started with a group of enthusiastic professionals, now working to satisfy its customers by adopting a unique, customer-centric style of consulting approach for delivering its products. Static data for Digital Twins is helpful. It represents the important details regarding the design and the specifications of an asset that would form the core of its digital replica. In general, the static data is essentially time-invariant but provides much detail regarding the structure and capabilities of an asset. For example, static data would refer to the design specifications, the properties of materials, and the blueprints of the structure of an asset that helps model the same asset with accuracy in a digital twin environment. Let's consider a Building Digital Twin as a case study. A digital twin refers to a detailed 3D virtual replica of a real property and platforms such as Matterport use AI to create those twins directly out of photos, producing very accurate and interactive models of real-world spaces. But what's interesting is that digital twins are not just static models – they can incorporate data from IoT, coming up with real-time and predictive insights. For instance, temperature or humidity level can be tracked with sensors, movement of employees followed with RFID, and vehicles or supplies with geolocation. This is an integration of static and dynamic data that allows a business to represent and study its assets in real time to respond with speed and make more well-informed decisions [6].

Digital twins are the biggest application of technologies in industries such as real estate, facilities management, and manufacturing in terms of value, throughout the lifecycle of the building or asset. They have great utility to AEC (architecture, engineering, and construction) and manufacturing teams when understanding and sharing spatial information across locations. Through static data with real-time input, digital twins contribute toward asset performance with a more global view. That is part of this evolution: **SimuTwin,** for example, creates software tools to simulate and optimize data flows, enable companies to model complex systems and processes that integrate data from different sources, and hence support informed decisions in sectors such as manufacturing, supply chain management, or urban planning. Another promising player is **Karpine Technologies Pvt Ltd**, incubated at the **AIC-NITTE Incubation Centre**, which provides blockchain- and IoT-based supply chain track-and-trace solutions to the fashion and luxury retail industries. They make use of digital twins to create an end-to-end traceability system emphasizing how the use of static data along with real-time tracking can drive better asset management as well as increased transparency across sectors. Indeed, it is imperative to point out the value that static data must bring to model an asset accurately and hence understand its capabilities as well as its limitations, and then, the integration of these streams of data with real-time data to provide actionable insights in digital twin systems.

3.2.4 Predictive Data: Looking into the Future

Having discussed what has already happened with historical data and what is going on with real-time data, we turn to the future with predictive data. Predictive Data is indeed a game changer. Predictive Data refers to the capability of predicting what will happen next. It is one kind of data developed by predictive analytics and simulation models of real-time scenarios in terms of predicting possible future states and behaviors of assets. In other words, if Historical Data gives one insight into the past and Real-Time Data captures the present, Predictive Data [7] can be considered as having a forecast of what lies ahead. Let us suppose a person owns a factory. We have a historical record of how machines have behaved over time, as well as real-time data that keeps you up to speed on current conditions. Knowing when a machine is probably going to fail, before it does, would be priceless. That's what predictive data does. It brings together all known information about an asset – past performance and present state – and feeds it through complex models to predict the potential problems that might be faced or when it will require maintenance [8].

For instance, data related to predictive maintenance is of high value for the avoidance of a surprise downtime. It ensures that maintenance happens only when necessary, minimizing the possibilities of unaccounted failures that will bring immense time and cost losses. Predictive maintenance [9] makes sure there's no frantic last-minute rush but smooth proactive upkeep. Besides preventing breakdowns, Predictive Data is a key driver in improving the efficiency of operations. The outputs from the simulation are useful in giving an indication of how an asset would behave under different conditions, hence optimizing the process. Predictive Data may indicate how an increase in speed on a production line could impact output or how changes in temperature or other environmental conditions might impact machinery.

Predictive Data also supports trend analysis. According to such analysis, some patterns would not have emerged by merely analyzing historical data or through real-time observation. These insights enable more effective decision-making in the allocation of resources, planning of production, and any possible expansion, keeping an organization ahead in a competitive environment. Overall, Predictive Data empowers proactive action rather than reactive action. It transforms Digital Twins into an effective decision-making tool that is more than just monitoring functions. Predictive data empowers organizations to prevent problems rather than just react to them when they happen. This forward-looking capability represents the core strength of a Digital Twin. For instance, a simulation has been created for RFID modules that predict access to hospital amenities, images, and codes available for reference as shown in Figure 3.1 and Figure 3.2.

*We simulated the RFID technology using **TKinter** library of Python and implemented machine learning tools (reinforced learning) to predict the availability of hospital amenities in real time.*

This is the simulation from which the data was collected in .csv format.

3.2.5 Dynamic Data

Dynamic Data means a constantly changing condition influencing physical assets. These are the kinds of definitions that characterize Dynamic Data – the type of information that requires constant updating. Dynamic Data represents in-stream sensor data, operations, and environment conditions all of which are continually changing. Since Historical Data is stuck in the past, while Predictive Data reaches into the future, it is the real-time update information that continuously maintains a Digital Twin as updated and valid and aligned directly with its actual current condition. An open-field-based wind turbine could see minute-by-minute fluctuations in the wind speed, temperature, and even vibrational levels. This data does more than monitor

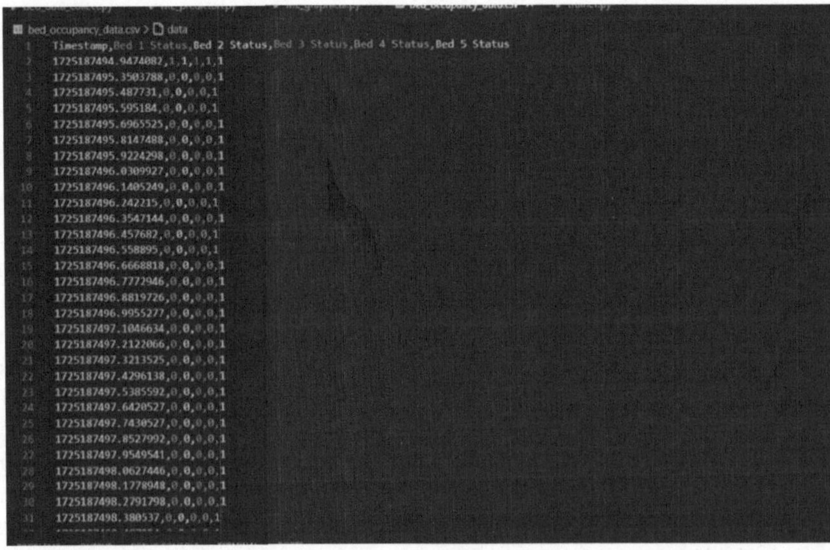
bed_occupancy_data.csv > data

```
Timestamp,Bed 1 Status,Bed 2 Status,Bed 3 Status,Bed 4 Status,Bed 5 Status
1725187494.9474082,1,1,1,1,1
1725187495.3503788,0,0,0,0,1
1725187495.487731,0,0,0,0,1
1725187495.595184,0,0,0,0,1
1725187495.6965525,0,0,0,0,1
1725187495.8147488,0,0,0,0,1
1725187495.9224298,0,0,0,0,1
1725187496.0309927,0,0,0,0,1
1725187496.1405249,0,0,0,0,1
1725187496.242215,0,0,0,0,1
1725187496.3547144,0,0,0,0,1
1725187496.457682,0,0,0,0,1
1725187496.558895,0,0,0,0,1
1725187496.6668818,0,0,0,0,1
1725187496.7772946,0,0,0,0,1
1725187496.8819726,0,0,0,0,1
1725187496.9955277,0,0,0,0,1
1725187497.1046634,0,0,0,0,1
1725187497.2122066,0,0,0,0,1
1725187497.3213525,0,0,0,0,1
1725187497.4296138,0,0,0,0,1
1725187497.5385592,0,0,0,0,1
1725187497.6420527,0,0,0,0,1
1725187497.7430527,0,0,0,0,1
1725187497.8527992,0,0,0,0,1
1725187497.9549541,0,0,0,0,1
1725187498.0627446,0,0,0,0,1
1725187498.1778948,0,0,0,0,1
1725187498.2791798,0,0,0,0,1
1725187498.380537,0,0,0,0,1
```

FIGURE 3.1 The actual simulation or Digital Twin of a hospital's room with five beds and shows on a scale of 0.00 to 1.00 which bed might be available

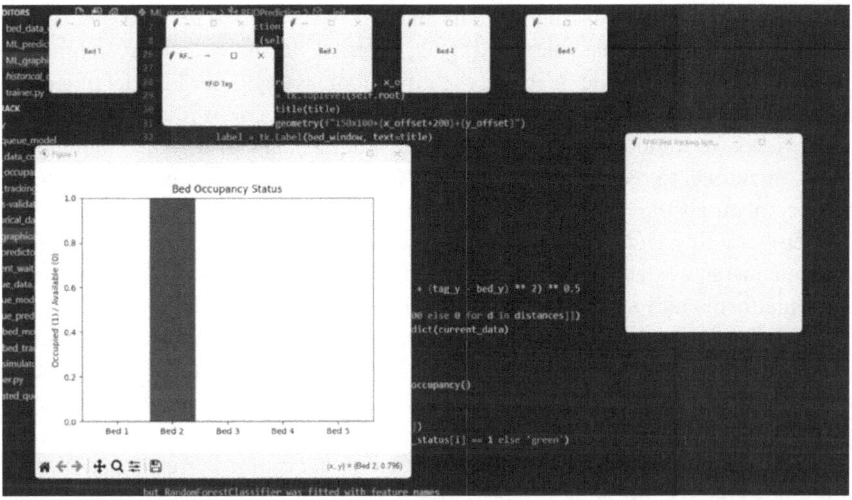

FIGURE 3.2 The actual simulation or Digital Twin of a hospital's room with five beds showing occupancy status

the present; it ensures that the Digital Twin stays abreast of the physical counterpart. Without it, the model would get outdated, like an aged GPS that still works but does not give accurate instructions, with a high potential to be misaligned in reality.

Now imagine it is a factory setting. The room's temperature may be volatile; machine speeds are constantly in fluctuation, and individuals never stop moving about inside. All these fall in the category of Dynamic Data, and all this will change dynamically in the Digital Twin so that the model represents the exact reality being there. This data flux in this dynamic environment might easily trigger real-time adjustments like adjustment in the machine's speed or adjusting the thermostat settings. What makes Dynamic Data truly value-added is its in-line integration into the Digital Twin, which ensures that the data remains relevant and timeline at all points. Dynamic Data ensures never to let the information turn obsolete to respond to the change in almost real-time situations, be it a temperature surge or a dip in performance. Dynamic Data constantly keeps the operators updated, ready to act against the happenings. Overall, Dynamic Data is what keeps a Digital Twin responsive and effective. Without it, the model would become static, and the ability to adapt swiftly would be lost. Dynamic Data captures ongoing reality in assets, so whatever is appearing in the Digital Twin mirrors events as they unfold.

3.2.6 Integrative Data

In the case of the Digital Twin, there are two prime dimensions, namely, Integration Data and Interoperability Data. Such data would ensure that a digital twin gives an all-round smooth view of the asset or system by drawing information from multiple sources and ensuring free flow between different systems. Integration Data pulls data from various sources and systems into a digital twin. The data source may be sensors, IoT equipment, manual inputs, external databases, and even feedback mechanisms. To be holistic in scope, the all-encompassing digital model of a physical system should include everything. Unless integrated, one would only have half-baked views from the digital twin. All the available data can then be combined by the organizations to receive a clearer, more holistic view of the asset – from design and specifications to real-time status in operational units [10].

Interoperability Data: This type of data focuses more on interoperability because of the inclusion of data within the digital twin-capability to transfer between different platforms and systems very easily. That is to say, it must provide the ability for different formats of data, protocols, and platforms to converse effectively with each other. Compatibility becomes a lifeline for the digital twin to function properly in complex environments where data needs to flow freely through legacy systems, modern platforms, or cloud services [11]. Interoperable systems make for real-time accurate insights in helping make better decisions. For example, **Nimaya Robotics Private Limited** is an illustration of integration and interoperability. It has breathed life into a Robotics Training System that aims to target people diagnosed with ASD along with other disabilities. Its system brings information from different hardware units enabled by IoT to record psychomotor and cognitive skills that are later analyzed in a cloud-based platform. Since the setup ensures seamless information exchange, these insights reach in real-time and accelerate up to 60%. Nimaya's solution puts out a lot, especially on how heterogeneous data sets and their interoperability can improve the efficiency of the digital system [8, 12].

3.2.7 User and Behavioral Data

Now that we have traveled this far on our expedition in this world of Digital Twin; there is another elephant in the room that gives us a solid foundation, let's shift gears to something as important – User and Behavioral Data. This kind of data does not come from the asset itself but rather from the people interacting with it. And we truly believe, understanding how users engage with your asset can be a game changer. The world today is becoming increasingly user centric – or at least involving people in some way. And whether it's an app on a smartphone, an appliance in a smart home, or even a car, how people use those assets provides incredible insight. That's where User and behavioral data come into play. This data tracks how the users interact with the asset, giving you a clear picture of their preferences, behaviors, and patterns.

In case of user interaction logs, behavioral analytics, and feedback, all constitute a treasure trove of information as far as people using the particular product are concerned. Such data shall answer critical questions like whether users are having a problem with those features, where there seems to be an imbalance in the functionality usage, or what's working and what's not. Let's take an example of a new car with fancy dashboards containing digital displays. User and Behavioral Data will let you know how people like operating touchscreens or like to have physical keys. Maybe the users are complaining that some controls are out of reach or hard to apply them.

This data gets analyzed for adjusting the design, enhancing functionality, and hence building up a better user experience [4].

The data is important and, specifically, for any consumer-facing product where user satisfaction is a key goal. However, it goes beyond making people happy. You're improving the general performance of the asset in as much as you are understanding how users behave. For instance, a better optimization of the ways by which people use and interact with a system may result in greater efficiencies and reduced wear and tear on the asset itself. It's not bound by one specific industry. Whether healthcare, automotive, or aerospace, knowing how people interact with your assets will always help you make improvements. An example would be a startup company called **TwinData Solutions**, focusing on managing data concerning Digital Twins as well as developing platforms that integrate all types of data, such as user interaction data, seamlessly. That means the Digital Twin will be kept current and reflects in real life how the asset is being used. Both User and Behavioral Data will help you continue to improve your Digital Twin by understanding the human part of the equation. And it's not just about the technical specs or the real-time data coming from sensors; it is about making the asset work better for the people that use it.

3.3 DATA COLLECTION SOURCES

There is no single place they are coming from; they are coming from all different sorts of sources, and all of them are playing their part in a bigger picture. All this makes for a true reflection of the real asset that the Digital Twin is going to be. First and foremost, we have sensors. These are basically basic devices that collect data on a good number of other physical parameters, such as temperature, pressure, and vibration. These can be termed as the "eyes" and "ears" of the system. From the car engine to a smart thermostat in your home, sensors are present all around. They constantly take data back, providing real-time information on how things function. Without sensors, we would have no idea what's going on at any given time within that asset. For example, if you're tracking a wind turbine, all of these temperature sensors, pressure sensors, and vibration sensors put all this together to monitor the condition of the turbine. And if things start going bad – if the temperature surges and goes wild – then there could be alerts created in those sensors, and it would be recognized right away.

And this brings us to the Internet of Things: IoT devices. These are a bit more advanced because instead of just capturing information, they are also communicating with the other systems. It's like a web of sensors and actuators. They feed back into the central system that enables seamlessness and allows for real-time monitoring right up to controlling. So, when using IoT devices, you are no longer passively collecting information [13]; you're actively managing the asset in real time. For instance, in a smart building, the IoT devices can automatically regulate air-conditioning based on information they get from temperature sensors installed throughout all areas, ensuring such a building will always function in an optimal way. In some instances, human input is necessary, and this will be achieved by introducing manual input. This is data that requires human intervention, like maintenance logs, inspection reports, or even feedback from the ground from the workers. Input by an engineer to fill in that gap when automated data collection isn't possible is valid. It's useful when you need more subjective information, like how something feels to operate, or if it feels off even though sensors don't find it. This data, in general, usually supports the autoref data and is found to offer greater detail and accuracy to the information.

Another such crucial source of information is related to external databases. These are databases that add an extra layer of context or information that might not be directly linked to the asset but impacts it in some fashion. Take weather data for example; these could be pretty critical when monitoring outdoor assets such as bridges or solar panels. Conditions can pretty much whip the performance upside their heads. You will also include market trends, regulatory information, or economic forecasts that support your decision-making on whether to keep or operate the asset in the external databases. Including all these external data [14] within the Digital Twin gives you a holistic model that represents all factors affecting the operation of your asset. Feedback data arise from the actual users or operators of your asset. Whether it is in the form of a feedback survey, statistics on usage data, or such basic interaction tracking data, feedback offers critical insight into the nature of the actual usage of an asset. Not knowing that the asset is performing technically is never enough; knowing how it is performing for those who use it is also critical. Feedback data helps bring out inconsistencies and issues that users might face, giving a clearer picture of the total experience. More relevant toward improving functionality is the fact that the asset not only functions properly but also well for the users. All of these sources – sensors, IoT devices, manual input, external databases, and feedback – come together to form the complete picture that a Digital Twin needs. Without any one of them, the model wouldn't be as accurate or effective. With all of them, it ensures that the Digital Twin is actually a real-time, living reflection of its physical counterpart, ready to provide insights, predict issues, and help in taking decisions.

3.4 DATA QUALITY AND CONTROL

Valid digital data constitutes the nerve supply of DTs while at the same time ensuring reliability in effectiveness. However, good data must come in consistent as well as secure to pass the integrity test: therefore, there is great room for debate on data of what's good for everyone and in what state. The first of the 3C's is accuracy. This element ensures that the data collected is correct and precise. Faulty sensors or flawed data collection processes lead to inaccurate data, and this can result in poor decisions. Sensor calibration and data validation techniques play a crucial role here. It is very important to implement error detection and correction processes; otherwise, building a Digital Twin would be like constructing on unstable ground, which could lead to flawed predictions and serious operational issues. Consistency is the next, whereby data is kept uniform across all sources and over time. It cuts out variations of discrepancies between data entries. Consistency depends on adequate integration of data with proper standardization protocols and regular audits, especially when consolidating data from different systems. This would align all data sources working without any pieces out of place, much like ensuring all puzzle pieces fit to make up a complete picture [15, 16].

Completeness refers to the accrual of every piece of required data, such that no significant information is left lacking. Amidst accuracy and consistency, partial data, therefore, fill in holes for the Digital Twin. Strategies for collecting comprehensive data, along with gap analysis, are essential for that purpose. All these endeavors ensure that no vital elements of data are not accounted for. Another dimension is timeliness. Real-time systems rely on receiving data right at the time needed for instant, informed decisions. Such delays in receiving data detract from the value information and may result in wasteful outcomes. Efficient pipelines and real-time processing frameworks have ensured that data remains "live" and current-enabling timely responses and judgments [17, 18].

Lastly, security is a must. When breaches of data are so prevalent, security cannot be optional in Digital Twins. Efforts to protect such data from theft and access should be quite strenuous. First lines include encryption, access controls, and storage solutions, among others. Third-party security audits and assessments help in exposing various vulnerabilities [5, 19–21] to keep the data as safe and secure as possible. Without these security measures put in place, the system of the Digital Twin would be vulnerable, and it will not take long before physical assets fall with them. Overall, high-quality data forms the basis of Digital Twins. Data needs to be accurate, complete, consistent, timely, and secure. Without them, even the most advanced digital twin will falter and become unreliable [4, 22].

3.5 CONCLUSIONS

The necessity of quality and control related to data in the paradigm of the digital twin is an important aspect. Proper working of digital twins is done only by accurate data that is complete and timely. Bad data may lead to bad models, which might therefore provide incorrect decisions, hence potentially leading to system failures. Sound data management practices, including sensor calibration, data validation, real-time processing, all help allow the digital twins approach to preserve the integrity of said systems. Additionally, data security will be required to protect sensitive information and ensure reliability in insights from digital twins. As more tools and integration technologies become available to drive the evolution of Digital Twins, clear, precise data requirements will be increasingly important to communicate. It has been noted in this chapter how good data structure is necessary for retrieval, processing, and data analysis in realizing the full potential of digital twins: placing them firmly at the heart of Industry 4.0. Such enhanced data management capabilities in digital twins owe much to innovations from startups in technology. As this chapter goes on to reveal, it is in fact a very large number of startups that design tools and platforms with the objective of optimizing the quality of data collection and analysis, respectively, for digital twin models. The ever-evolving cloud-based platforms that want to simplify the validation and processing of data keep developing vigorously yet focus on real-time data management companies that may help build ecosystems around the digital twin aimed at improving far more advanced, reliable, and trustworthy models. Innovations as such are of huge significance for industries like aerospace, defense, and automotive, as well as smart cities, which are merely at a nascent stage of using digital twins to manage such high-value complex assets. Without data, Digital Twins would represent nothing more than a fancy idea with no potential to revolutionize an industry. All industries from new startups to the largest industries – aerospace, manufacturing, or agriculture – use Digital Twins to make better, faster decisions. Whether it's ensuring that factories stay in operation for as much time as possible without downtime, optimizing performance levels, or even predicting eventual failures before they happen, Digital Twins are changing how we understand and operate physical assets.

REFERENCES

1. Jones, D., Snider, C., Nassehi, A., Yon, J., & Hicks, B. (2020). Characterising the Digital Twin: A Systematic Literature Review. *CIRP Journal of Manufacturing Science and Technology*, 29, 36–52. https://doi.org/10.1016/j.cirpj.2020.02.001
2. Tao, F., Zhang, M., Liu, Y., & Nee, A. Y. C. (2019). Digital Twin in Industry: State-of-the-Art. *Annual Reviews in Control*, 47, 5–20. https://doi.org/10.1016/j.arcontrol.2019.01.001
3. Sharma, L., & Carpenter, M. (Eds.). (2022). *Computer Vision and Internet of Things: Technologies and Applications* (1st ed.). Chapman and Hall/CRC. https://doi.org/10.1201/9781003244165
4. Qi, Q., & Tao, F. (2018). Digital Twin and Big Data Towards Smart Manufacturing and Industry 4.0: 360 Degree Comparison. *IEEE Access*, 6, 3585–3593. https://doi.org/10.1109/ACCESS.2018.2793265
5. Kritzinger, W., Karner, M., Traar, G., Henjes, J., & Sihn, W. (2018). Digital Twin in Manufacturing: A Categorical Literature Review and Classification. *IFAC-PapersOnLine*, 51(11), 1016–1022. https://doi.org/10.1016/j.ifacol.2018.08.474
6. Negri, E., Fumagalli, L., & Macchi, M. (2017). A Review of the Roles of Digital Twin in CPS-based Production Systems. *Procedia Manufacturing*, 11, 939–948. https://doi.org/10.1016/j.promfg.2017.07.198
7. Føre, M., Alver, M. O., Alfredsen, J. A., Rasheed, A., Hukkelås, T., Bjelland, H. V., Su, B., Ohrem, S. J., Kelasidi, E., Norton, T., & Papandroulakis, N. (2024)., Digital Twins in intensive aquaculture—Challenges, opportunities and future prospects. *Computers and Electronics in Agriculture*, 218, 108676. https://doi.org/10.1016/j.compag.2024.108676
8. Dammacco, L., Carli, R., Lazazzera, V., Fiorentino, M., & Dotoli, M. (2022). Designing complex manufacturing systems by virtual reality: A novel approach and its application to the virtual commissioning of a production line. *Computers in Industry*, 143, 103761. https://doi.org/10.1016/j.compind.2022.103761
9. Dahir, H., Luna, J., Khattab, A., Abrougui, K., & Kumar, R. (2023). Chapter 4 - Challenges of Digital Twin in healthcare. In Abdulmotaleb El Saddik (editor), *Digital Twin for Healthcare*, Academic Press, pp. 73–95. https://doi.org/10.1016/B978-0-32-399163-6.00009-3
10. Boyes, H. & Watson, T. (2022). Digital twins: An analysis framework and open issues. *Computers in Industry*, 143, 103763. https://doi.org/10.1016/j.compind.2022.103763
11. Carpenter, M., Sharma, L., Sengupta, S., & Kumar, A. (2023). A Vision-Based System in Healthcare: Use Cases and Real-Time Applications. In L. Sharma & P. K. Garg (Eds.), *Technological Prospects and Social Applications of Society 5.0* (pp. 17–34). CRC Press. https://doi.org/10.1201/9781003324720-8

12. Sangari, M. S. & Mashatan, A. (2022). A data-driven, comparative review of the academic literature and news media on blockchain-enabled supply chain management: Trends, gaps, and research needs. *Computers in Industry*, 143, 103769. https://doi.org/10.1016/j.compind.2022.103769
13. Glaessgen, E., & Stargel, D. (2012). The Digital Twin Paradigm for Future NASA and U.S. Air Force Vehicles. In *53rd AIAA/ASME/ASCE/AHS/ASC Structures, Structural Dynamics and Materials Conference 20th AIAA/ASME/AHS Adaptive Structures Conference 14th AIAA*, p. 1818. https://doi.org/10.2514/6.2012-1818
14. Niloofar, P., Francis, D. P., Lazarova-Molnar, S., Vulpe, A., Vochin, M. C., Suciu, G., Balanescu, M., Anestis, V., & Bartzanas, T. (2021). Data-driven decision support in livestock farming for improved animal health, welfare and greenhouse gas emissions: Overview and challenges. *Computers and Electronics in Agriculture*, 190, 106406. https://doi.org/10.1016/j.compag.2021.106406
15. Sharma, L., & Carpenter, M. (Eds.). (2022). *Computer Vision and Internet of Things: Technologies and Applications* (1st ed.). Chapman and Hall/CRC. https://doi.org/10.1201/9781003244165
16. Fuller, A., Fan, Z., Day, C., & Barlow, C. (2020). Digital Twin: Enabling Technologies, Challenges, and Open Research. *IEEE Access*, 8, 108952–108971. https://doi.org/10.1109/ACCESS.2020.2998358
17. Sharma, L., & Garg, P.K. (Eds.). (2024). *Deep Learning in Internet of Things for Next Generation Healthcare* (1st ed.). Chapman and Hall/CRC. https://doi.org/10.1201/9781003451846
18. Lu, Y., Liu, C., Kevin, I., Wang, K., Huang, H., & Xu, X. (2020). Digital Twin-driven smart manufacturing: Connotation, reference model, applications and research issues. *Robotics and Computer-Integrated Manufacturing*, 61, 101837. https://doi.org/10.1016/j.rcim.2019.101837
19. Pushpa, J., & Kalyani, S. A. (2020). Chapter 3 - Using fog computing/edge computing to leverage Digital Twin. In Pethuru Raj and Preetha Evangeline (editors), *Advances in Computers*, Elsevier, Vol. 117, Issue 1, pp. 51–77. https://doi.org/10.1016/bs.adcom.2019.09.003
20. Augustine, P. (2020). Chapter 4 - The industry use cases for the Digital Twin idea. In: Pethuru Raj and Preetha Evangeline (editors), *Advances in Computers*, Elsevier, Vol. 117, Issue 1, pp. 79–105. https://doi.org/10.1016/bs.adcom.2019.10.008
21. Sharma, L., & Garg, P.K. (Eds.). (2021). *Artificial Intelligence: Technologies, Applications, and Challenges* (1st ed.). Chapman and Hall/CRC. https://doi.org/10.1201/9781003140351
22. Sharma, L., Carpenter, M., & Sengupta, S. (2024). Improved Patient Care Using Robotics in the Healthcare Industry. In L. Sharma & P. K. Garg (Eds.), *Deep Learning in Internet of Things for Next Generation Healthcare* (pp. 164–189). CRC Press. https://doi.org/10.1201/9781003451846

4 A Digital Twinning for Advanced Precision Medicine and Patient Care

Tools and Technologies

Sudhriti Sengupta, Lavanya Sharma, Abhishek Kumar, and S B Goyal

4.1 INTRODUCTION

A digital twin can be said to be digitized, virtualized, and modeled equivalent of a real entity. Digital twins (DTs) are capable of producing visual representations of the dynamics and components of a gadget. A virtual depiction of a real-world system, process, or physical thing that replicates it in real time is called a digital twin [1]. In 1997, the term "digital twin" was first instigated in research work done by Hernández and Hernández [2]. This technology generates a dynamic model that can be examined and altered by utilizing data from sensors and other sources. In order to monitor and model operations and maximize efficiency, DTs are extensively utilized in the engineering and industrial sectors. Digital twins were first used in the aerospace sector, most notably by NASA during the Apollo space program. One of the two identical space shuttles that NASA built is in orbit, and the other is its digital twin on Earth. Users and stakeholders can now obtain thorough insights into the inner workings of almost any system, including the interconnections between its constituent parts and forecasts of its future behavior, due to technological advancements. Digital twins are becoming more and more common in a variety of industries, including engineering, healthcare, and freight management. Digital twins of patients can help in personalized medicine, allowing for tailored treatment plans based on real-time health data. The patient, in actual life, can be referred to as the physician entity in the context of medicine. The patient is represented digitally in virtual space by a virtual entity. Because these digital copies share traits with the patient, data gathered from the patient can be used to forecast and simulate biological processes or disease states. To enable constant updating of the parameters that represent the physical entity's state, the physical and virtual entities interact through physical-to-virtual connections. A

DOI: 10.1201/9781003582489-4

medical DT is essentially a virtual environment in which clinical and medical judgments can be tested prior to being implemented in a real patient. Medical DT is essentially a virtual environment in which clinical and medical judgments can be tested prior to being implemented in real patients [3]. The digital twin facilitates the perception of the subterranean environment, which greatly advances the development of green and sustainable coal resource exploitation [4]. Digital twins represent a transformative approach to mineral exploration, enabling companies to harness data and simulations to make informed decisions, reduce costs, and minimize environmental impacts. The COVID-19 epidemic drove the market's growth potential by boosting the expansion of the digital twin market across a range of sectors, including retail, healthcare, energy, and real estate. As a result, it is anticipated that a number of countries will incorporate Digital Twin solutions into their economic reform initiatives. Similarly, a number of enterprises are implementing Digital Twin technologies to streamline their supply chains and operating procedures in an effort to recover from the economic disruptions brought on by the pandemic [5]. In manufacturing, DT is used to optimize production processes, improve efficiency, and reduce downtime by predicting equipment failures. It allows manufacturers to create virtual prototypes of products. This enables testing and refinement of designs before physical production, reducing time-to-market and costs. Digital twins offer transformative potential across diverse sectors by integrating real-time data with simulation capabilities. As technology advances, the scope and effectiveness of digital twins are expected to expand, driving innovation and efficiency in numerous applications, like, agriculture, construction and real estate, Telecommunication, Energy Sector, Urban planning, and many more. It takes a considerable amount of effort to achieve the full potential of digital twins [6]. The introduction of Industry 4.0 placed a strong emphasis on both the digitization of data and the incorporation of sensors into industrial environments, which is largely responsible for the development of digital twins. One of the essential applications of digital twins is pertaining to healthcare and medical facilities. There are numerous applications for digital twin technology in the medical field. The healthcare and life sciences sectors were compelled to accelerate their digital transformation initiatives in order to facilitate the widespread transition to the pandemic. Digital healthcare procedures must become more effective. This chapter aims to provide insights into improved patient care with the advent of digital twin technology. Section 4.2 discusses the core components of DT, and the technologies used in DT are discussed in Section 4.3. The application of DT in different cores is discussed in Section 4.4. Section 4.5 elaborates on the application, components, and challenges of DT in patient care. This is followed by a conclusion in Section 4.6.

4.2 COMPONENTS OF DIGITAL TWINS

Digital twins consist of several key components that work together to create a dynamic, virtual representation of physical objects, systems, or processes. In DT, virtual representations of physical things during their lifecycle can be understood, directed, and reasoned with real-time data or a simulation model that collects data from the field and initiates the functioning of physical equipment. A digital twin is a real-time digital depiction of a physical thing, according to Fu et al. [7]. They offer detailed representations of actual items and are connected to them remotely. They include dynamic behavior in addition to static product designs like CAD models as shown in Figure 4.1.

In 2014, Grieves observed that DT has three principal components, viz., physical object, digital world, and association existing between the physical and digital world [8, 9]. Resources, equipment, objects, specific regulations, end users, and environment are some of the elements that comprise the complex, dynamic, industrial environment known as the physical world. The physical object consists of two main parts: Devices and Sensors. Devices represent real physical objects that are modeled to their virtual "twin" counterparts. Sensors are physically connected electrical components that act as the source of data and information. Among the most often used sensors are radio frequency identification, rapid response codes, and programmable logic controllers (PLCs). The digital world consists of a Virtual Environment Platform (VMP), which creates a complete and integrated 3D simulation for implementation and applications and enables functions to demonstrate how various algorithms work. For the creation and implementation of the DTs, VMP offers a number of crucial schemes [10]. The connection between the physical object and the digital world forms an important component of the Digital twin. It depends on the method adopted by the developers. These connections are called middleware in DT terminology. Bakken [11] defines middleware as a collection of techniques designed to facilitate communication between systems that exhibit heterogeneity and complicated properties.

4.3 TECHNOLOGIES USED IN DIGITAL TWINS

Digital Twins are virtual representations of physical systems, assets, or processes integrating a variety of technologies to simulate, monitor, and analyze real-world

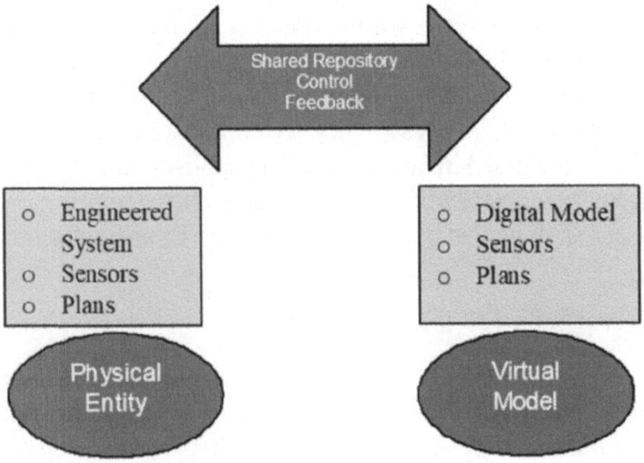

FIGURE 4.1 Component of digital system

scenarios. The three primary elements of Digital Twins are data acquisition, data simulation, and data usage as shown in Figure 4.2. Digital Twin employs four technologies to gather and save data in real time and extract information to offer valuable insights.

A) ***Internet of Things (IoT):*** IoT devices provide real-time data from the physical world, such as temperature, pressure, humidity, speed, or energy consumption. IoT is the primary technology used by digital twins in all of their applications. Over 90% of all IoT platforms will be capable of digital twinning by 2027 [12]. Sensors are used by IoT to gather data from actual items. A digital clone of a physical thing is made using the data that the Internet of Things transmits. After that, the digital version can be examined, altered, and improved. IoT helps Digital Twin applications build a real-time virtual representation of a physical object and continuously updates data. IoT systems aggregate data from multiple sources, feeding it into the digital twin for accurate, up-to-date simulations. As a consequence, IoT is the main technology used in all Digital Twin applications.
B) ***Cloud Computing*****:** Delivering hosted services over the Internet is referred to as cloud computing. The technology effectively uses the Internet to store and retrieve data. The cloud offers scalable infrastructure for storing and processing vast amounts of data needed for digital twins. With the use of cloud computing, Digital Twins with massive data sets can store their data in a virtual cloud and quickly retrieve the information they need from anywhere. Cloud-based platforms enable real-time collaboration and accessibility across multiple locations and teams. Cloud computing helps Digital

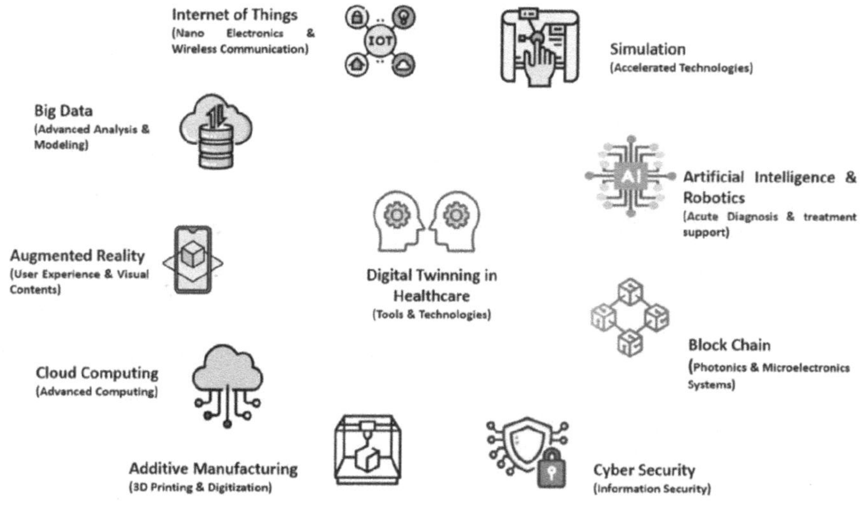

FIGURE 4.2 The four technologies that comprise the functional and operational unit of Digital Twins

Twins to successfully decrease the processing time of complicated systems and overcome the challenges of retaining vast volumes of data.

C) ***Artificial Intelligence (AI)*:** Artificial Intelligence (AI) is a field of computer science focused on creating systems that can perform tasks that typically require human intelligence. This includes capabilities like problem-solving, understanding language, recognizing patterns, and even simulating aspects of human perception and decision-making. AI models can forecast potential issues and suggest improvements based on historical data and real-time input. Artificial Intelligence (AI) can assist Digital Twins by offering a sophisticated analytical tool that can automatically analyze collected data and produce insightful information, forecast results, and offer advice on how to prevent potential challenges [13].

D) ***Extended Reality (XR):*** Extended Reality (XR) is an umbrella term that encompasses all real and virtual combined environments, blending physical and digital worlds to create immersive experiences. XR includes three primary technologies: Virtual Reality (VR), Augmented Reality (AR), and Mixed Reality (MR). Each provides unique ways for users to interact with and experience digital content in various degrees of immersion. AR and VR allow users to interact with the digital twin in immersive environments, enhancing understanding of complex systems. Technicians can use AR/VR to perform maintenance tasks virtually, seeing real-time data overlaid on physical systems [14].

4.4 APPLICATION OF DIGITAL TWINS

Digital twins have applications across numerous industries, as they provide a dynamic and data-driven replica of physical assets or systems, enabling real-time monitoring, analysis, and predictive insights. The DT is currently possibly used in a variety of industries that require big data analytics, cloud computing, the Internet, and other technologies because of their ongoing integration and mapping between the physical as well as virtual environments as shown in Figure 4.3. Digital twins are now mostly used in engineering and manufacturing to create precise virtual objects and operational process simulations. Some of the widely popular areas of Digital twins are discussed below:

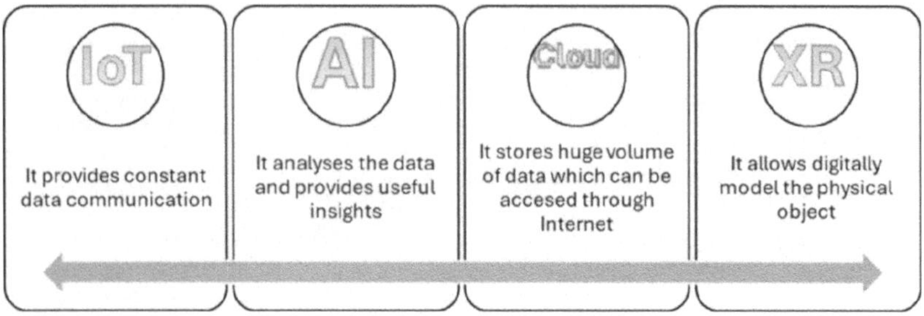

FIGURE 4.3 Empowering technologies in healthcare systems

A) **In manufacturing,** digital twins are transforming production, operations, and asset management by creating real-time, data-driven virtual representations of equipment, processes, or entire manufacturing facilities. Digital twins are integral to the evolution of the modern manufacturing industry, enabling smarter, more efficient, and sustainable production. Through real-time insights, predictive analytics, and enhanced operational control, digital twins are helping manufacturers move toward fully autonomous, smart factories aligned with Industry 4.0 principles [15].

B) **The modernization and development of cities**, as well as the establishment of smart cities, require the introduction of intelligent systems, technologies, and solutions in domains other than transportation. Digital twins are revolutionizing smart cities by creating virtual representations of urban environments that allow city planners, administrators, and residents to interact with and analyze real-time data for optimized city management. Digital twins are foundational to the concept of smart cities, offering a comprehensive, data-driven approach to urban management and planning. By integrating IoT, AI, and real-time analytics, digital twins empower cities to become more adaptive, resilient, and sustainable in the face of modern challenges [16] as shown in Figure 4.3.

C) Every **economy** depends on the agriculture sector for its continued operation. In addition to being a significant supplier of food and raw materials, this business is also a major employer of people in the community. Food security, sustainability, productivity, and profitability are becoming more pressing issues as the world's population grows and the market demands higher product quantities and quality requirements. In agriculture, digital twins are transforming traditional farming practices by creating virtual replicas of farms, fields, equipment, and even individual crops or animals. These digital models allow farmers to monitor, simulate, and optimize agricultural operations in real time, driving greater efficiency, sustainability, and productivity. As digital twins continue to evolve with AI, IoT, and big data advancements, their applications in agriculture will become even more powerful. Future developments may include hyper-personalized crop models, real-time market analysis, and advanced integrations with AI for even more sophisticated prediction and optimization. Overall, digital twins promise to make agriculture smarter, more sustainable, and more productive, helping to meet the demands of a growing global population [17].

D) **Automotive and Transportation:** Digital twins play a transformative role in the automotive and transportation sectors, enhancing design, production, maintenance, and overall user experience. Engineers use digital twins to create virtual models of vehicles, allowing for testing and optimization before physical prototypes are built. Digital twins simulate vehicle performance under various conditions, helping to identify design flaws and improve efficiency. Digital twins can analyze driver behavior and preferences, enabling personalized features and improving user satisfaction. They can optimize in-vehicle systems (e.g., navigation, entertainment) based on real-time data and user interactions. The application of digital twins in the automotive and transportation sectors leads to increased efficiency, safety, and innovation. As technology continues to evolve, the potential for digital twins to shape the future of mobility will only grow, paving the way for smarter vehicles and more efficient transportation systems [18].

The DT is currently possibly applied in various domains that include organizing, cooperation, and integration among both the physical and virtual spaces as an outcome of ongoing collaboration with the Internet, big data analytics, cloud computing, and other technologies. Applications for DT can be found in a variety of industries, including manufacturing, electricity, smart cities, healthcare, freight shipping, construction, automobiles, drilling platforms, etc. The usage of digital twins spans a wide range of applications, each tailored to specific industry needs. As technology advances, the capabilities and applications of digital twins will continue to expand, further driving innovation and efficiency. Digital twins provide promising insights into patient care and represent a cutting-edge approach to personalized medicine and healthcare management. By creating virtual models of patients, healthcare providers can enhance diagnosis, treatment, and overall patient outcomes. Digital twins are reshaping patient care by creating virtual models of patients that allow healthcare providers to monitor, predict, and personalize treatment based on real-time data. The following sections discuss the role and usage of Digital twins that are adaptable to patient care and health support systems.

4.5 ROLE AND SCOPE OF DIGITAL TWINS IN PATIENT CARE

Digital twins in patient care represent a cutting-edge approach to personalized medicine and healthcare management. These DTs offer dynamic health guidance, improve disease treatment options, and allow a quantitative examination of critical life processes using a variety of health parameters as shown in Figure 4.4. This novel

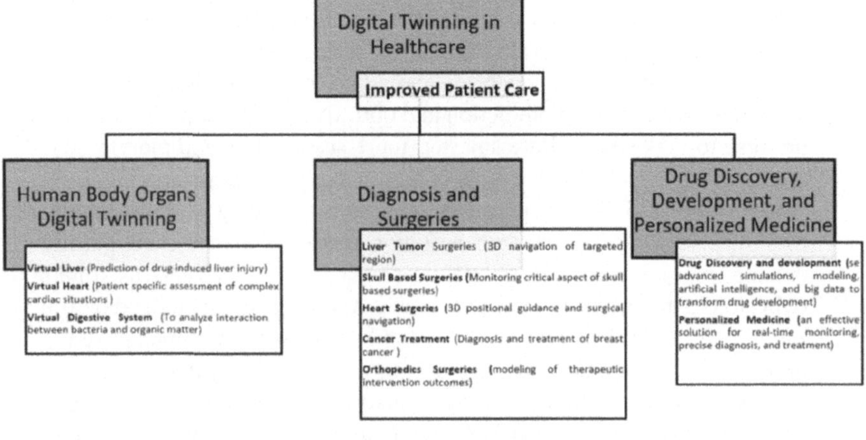

FIGURE 4.4 Digital twinning in improved patient care

strategy seeks to improve our mathematical knowledge of biological processes, revolutionize current therapeutic procedures, and genuinely customize healthcare. In this section of the chapter, the role and application of Digital twins in Patient care, its main components, challenges, and a framework for application of DT are discussed.

4.5.1 Application of Digital Twins in Patient Care

The following sections look shows some of the core areas of usage of digital twins in patient care.

A. **Monitoring of Individual Physiological Parameters**

Every individual has a slightly distinct classification of normal or healthy states based on experimentation or clinical definitions, hence these classifications cannot be applied to expansive populations. At present, low-resolution data and a summarized picture of a patient's clinical history are used for treatment individualization. For instance, the definition of normal blood pressure is still unclear. The comparatively few blood pressure readings and the failure to evaluate the influence of physiological and behavioral patterns in any one person could be the causes. It is challenging to identify departures from normal that eventually make up the disease state in the absence of a tailored definition of normal. Through constant information exchange between the patient and their DT, a medical DT can define normal in each individual. Disease is defined by deviations from this normal condition, and intervention outcomes can be predicted using these deviations [19].

B. **Development of New Drugs and Medications**

Testing and monitoring to determine the safety and efficacy of new medications can be achieved using Digital Twin technology. Massive volumes of data are generated at every stage of the drug development process, which need to be managed. These data are used by digital twins to build a model. Digital twins can thereby expedite drug research clinical trials, enabling them to proceed more swiftly and requiring fewer patients to be assigned to receive medication. Digital twins can be used to simulate patient populations for clinical trials, potentially accelerating the development of new treatments and therapies. Researchers can analyze aggregated data from digital twins to discover new trends and insights in patient care.

C. **Cancer Management**

By merging clinical and individual proteome data with population data, medical DTs can also fulfill the potential of precision oncology. These models can be used to modify therapies in real time and continually gain insight from fresh data as well as unique patient care decisions made by physicians. This is especially crucial in cases of medication resistance or cancer recurrence, as patients may need different radiation, chemotherapy, or surgery regimens based on their innate resistance. Based on the genome of each patient's tumor, medical DT platforms can provide alternate treatment plans and anticipate when resistance will start. In order to reduce harmful side effects, chemotherapy regimens can also be tailored to the patient's metabolism. These models have demonstrated potential in forecasting triple-negative breast cancer treatment outcomes [20].

D. **Training and Education**

For a while, virtual reality (VR) technology has been employed in medical education to enhance and augment conventional health education. Additionally, the technology assisted surgeons in performing robotic surgery and physicians in treating patients. With the help of augmented and virtual reality, artificial intelligence (AI)-enabled digital twins can help with the training and education requirements of medical professionals. To encourage interactive learning and reduce the use of cadavers, for instance, many businesses have implemented medical anatomy and surgical procedure simulations. Medical schools can teach surgical methods in simulated forms by using Digital Twin technologies [21].

E. **Management of Cardiovascular Diseases**

Effective acute care and guidelines-based risk factor management techniques improve survival and quality of life in cardiovascular disease. Traditional simulation models and precursor models at various scales can be used to build DTs, which can offer customized treatments. For instance, an inverse analytic DT system can use neural networks to identify abdominal aortic aneurysms and provide severity values [22]. By implanting virtual electrodes, the Siemens Digital Heart is used to assess the effectiveness of cardiac resynchronization therapy.

F. **Surgical Practice**

The concept behind a surgical digital twin is that a patient model is made, and surgery can be scheduled during a multidisciplinary team conference, practiced in advance in a simulator, and consulted during the procedure to confirm anatomy and prevent unintentional structural damage. In order to reduce patient risk, this real-time patient model may potentially lead to clinical trials in which novel tools, methods, or treatments are initially tested on the digital twin. By enabling simulated practice in the context of each patient's unique anatomical and physiological variation, digital twins and the rapidly evolving virtual reality platforms can also improve surgical training for practitioners [23, 26, 27, 28].

G. **Hospital Management**

Hospitals, capacity planning, workflows, staffing systems, and care delivery systems can all be optimized with the use of digital twins. By watching patient flow internally and predicting possible surges using external data, Digital Twins, for instance, might improve predictive analytics and lessen the burden of erroneous bed occupancy projections. Healthcare providers can improve patient care, operational effectiveness, and profitability by using the data from Digital Twin solutions to think more strategically about capacity and resources based on improved forecasting. Digital twin solutions could save costs and enhance a patient's experience in a hospital, which is a challenge for hospitals trying to save operational expenses and stay competitive [29, 30].

4.5.2 Components of Digital Twins in Patient Care

With the integration of DT technologies, healthcare is poised for a dramatic transformation in an era characterized by groundbreaking technology advancements [31, 32, 33]. Five essential hallmarks, as discussed below, define a medical DT platform that embodies this paradigm shift toward individualized, data-driven healthcare as shown in Figure 4.5. The essential pillars of digital twins in healthcare are discussed below:

i. **Simulation:** It is a main component of DT, which are virtual models that accurately mirror physical, products, objects, or processes. DT can run various simulations to study various processes.
ii. **Metaverse Representation:** The prospect of a metaverse, a virtual environment of massive, realistic digital models and creatures that can be utilized

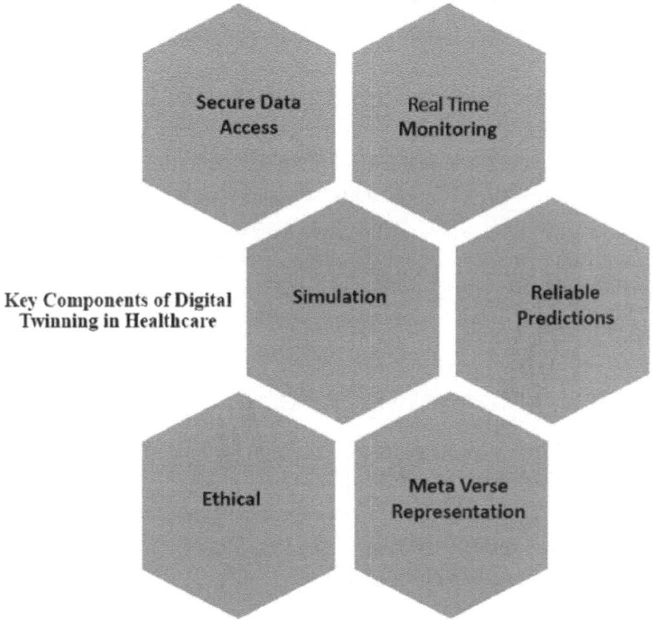

FIGURE 4.5 Key components of Digital Twin in healthcare

to modify treatment, track lifestyle changes, and assess response, is at the core of a medical DT platform.

iii. **Reliable Prediction**: Distinctive feature of a medical DT platform consists of AI agents. They are the platform's virtual brains, continuously sifting through enormous volumes of data to provide accurate forecasts and insights. These AI agents use extensive metaverse information, such as past medical records, treatment results, and patient-specific factors. They can forecast the course of a patient's health, the efficacy of certain therapies, and the likelihood of developing particular illnesses by modeling various scenarios. By taking a proactive stance, healthcare becomes predictive rather than reactive, allowing for early intervention and better treatment plans.

iv. **Real-Time Monitoring:** Real-time physical entity monitoring is another distinguishing feature of a medical DT platform. The metaverse's patient-specific avatars closely resemble the actual health conditions of their subjects. They are able to keep close tabs on the results of treatment, physiological factors, and vital signs. Patients are provided with continuous utilization of their health data due to this real-time monitoring, which guarantees prompt actions and a rapid response to anomalies.

v. **Secure Data Access:** A significant emphasis on secure data access, where patient data is protected with the greatest privacy and encryption standards, is another distinguishing feature of a medical DT platform. Healthcare providers are given secure, role-based access to pertinent patient data, and patients have preference over who can access their data. This secure data access adheres to strict data protection regulations and protects patient privacy.

vi. **Ethics:** One of the most important characteristics of a medical DT platform is ethics. It promotes the idea that any advancement in healthcare must be in the best interests of the patients. This platform complies with the strictest ethical guidelines for research, data use, and service delivery. It guarantees that patient rights are upheld and that consent is always obtained. The use of medical data to enhance patient care and the larger healthcare community while upholding openness and confidence is also part of the ethical commitment.

4.5.3 Challenges in Digital Twins in Patient Care

While digital twins have a lot of potential to improve patient care, using them in the healthcare industry presents a number of challenges and constraints [34, 35, 36].

- Data Acquisition and Integration
- Model Accuracy and Validation
- Computational Complexity and Resource Requirements
- Ethical and Privacy Concerns
- Cost and Accessibility

Integrating data from several sources, such as wearable sensors, medical imaging, genomic data, and electronic health records (EHRs), is necessary to produce precise and comprehensive digital twins. Uneven data formats, data interoperability challenges, and the possibility of erroneous or missing data make this task challenging. It is also crucial to guarantee data security and privacy when compiling this sensitive data. The reliability of a digital twin's forecasts and recommendations is directly impacted by its accuracy. It can be challenging to create models that properly capture the physiological processes and response to treatments of specific patients this requires complex algorithms and validation against actual results. Misdiagnosis or unsuccessful treatment plans may result from the model's oversimplification or errors. It requires an enormous amount of processing power and memory to simulate intricate physiological processes. Large amounts of real-time data from many sources can be computationally demanding to process and analyze, requiring high-performance computing hardware that not all healthcare facilities can afford or conveniently access. There are several significant privacy and ethical issues with using patient data to build digital twins. Ensuring data security and safeguarding patient confidentiality are essential for upholding confidence and adhering to laws such as Health Insurance Portability and Accountability Act (HIPAA). Particular attention must also be given to issues of data ownership, permission, and potential biases in algorithms. Digital twin technology can be costly to create and deploy, which could lead to unequal access to this cutting-edge, individualized care. Its adoption by less extensive healthcare practitioners or those serving underprivileged populations may be limited by the expense of technology, software, data storage, and specialist knowledge. Addressing these challenges will be crucial for realizing the full potential of digital twins in revolutionizing patient care. These challenges highlight the need for careful consideration and planning when implementing digital twins in patient care Collaboration between healthcare providers, technology developers, researchers, and policymakers is essential to overcome these obstacles and ensure the responsible and ethical implementation of this promising technology

4.6 CONCLUSION

The seamless data integration in each direction between a virtual and real machine is the concept underlying a digital twin. Various perspectives and approaches are utilized in various application domains and industries. According to a comprehensive review, digital twins have changed significantly since their conception and are at a stage where researchers are investigating useful technologies in current industries. Digital Twin technologies are growing rapidly to offer a highly advanced digital revolution to improve the world for humanity, despite obstacles in addressing data security and quality, rising power and storage demands, and integration with current infrastructures. Digital Twins will be extended to additional sectors and application cases in the future. The solutions will integrate additional technologies, like artificial intelligence (AI) for improved connections, insights, and analytics, and augmented reality (AR) for an immersive experience.

REFERENCES

1. S. Manickam, L. Yarlagadda, S. P. Gopalan and C. L. Chowdhary, "Unlocking the potential of digital twins: A comprehensive review of concepts, frameworks, and industrial applications," *IEEE Access*, 11, 135147–135158, 2023. doi: 10.1109/ACCESS.2023.3338530.
2. L. A. Hernández and S. Hernández, "Application of digital 3D models on urban planning and highway design," *WIT Transactions on Ecology and the Environment*, 30, 391–402, 1997.
3. K. Zhang, H. Zhou, T. Daniel, et al., "Concepts and applications of digital twins in healthcare and medicine" *Patterns*, 5(8).
4. Z. Xing, S. Zhao, W. Guo, F. Meng, X. Guo, S. Wang, and H. He, "Coal resources under carbon peak: Segmentation of massive laser point clouds for coal mining in underground dusty environments using integrated graph deep learning model," *Energy*, 285, 128771, 2023.
5. Technavio, "Digital twin market by end-user, deployment, and geography - Forecast and analysis 2021–2025," 2022. Available at: https://finance.yahoo.com/news/digital-twin-market-size-grow-154500181.html
6. F. Tao and Q. Qi, "Make more digital twins," *Nature*, 573(7775), 490–491, 2019.
7. Y. Fu, G. Zhu, M. Zhu, et al., "Digital twin for integration of design-manufacturing-maintenance: An overview," *Chinese Journal of Mechanical Engineering*, 35(80), 2022. https://doi.org/10.1186/s10033-022-00760-x.
8. M. W. Grieves, "Digital twin: Manufacturing excellence through virtual factory replication," in *Virtually Perfect: Driving Innovative and Lean Products Through Product Lifecycle Management*, p. 133, 2014.
9. S. Răileanu, T. Borangiu, N. Ivanescu, O. Morariu, and F. Anton, "Integrating the digital twin of a shop floor conveyor in the manufacturing control system," in *Proc. Int. Workshop Service Orientation Holonic Multi-Agent Manuf.*, Springer, pp. 134–145, 2019.
10. Y. Zheng, S. Yang, and H. Cheng, "An application framework of digital twin and its case study," *Journal of Ambient Intelligence and Humanized Computing*, 10(3), 1141–1153, 2019.
11. D. Bakken, "Middleware," 2003. Available at: https://eecs.wsu. edu/~bakken/middleware.pdf
12. Researchandmarkets, "Digital twins market by technology, twinning type, cyberto-physical solutions, use cases and applications in industry verticals 2022–2027," 2022, Available at: https://www.researchandmarkets.com/reports/5308850/digitaltwins-market-by-technology-twinning
13. Z. Lv, and S. Xie, "Artificial intelligence in the digital twins: State of the art, challenges, and future research topics," *Digital Twin*, 1(1), 12, 2021. https://doi.org/10.12688/digitaltwin.17524.1
14. B. Marr, "What is extended reality technology? A simple explanation for anyone." Aug. 12, 2019. Available at: https://www.forbes.com/sites/bernardmarr/2019/08/12/
15. M. Soori, B. Arezoo, and R. Dastres, "Digital Twin for Smart Manufacturing: A Review," *Sustainable Manufacturing and Service Economics*, 2, 100017. https://doi.org/10.1016/j.smse.2023.100017.
16. L. Deren, Y. Wenbo, and Z. Shao, "Smart city based on digital twins," *Computational Urban Science*, 1(4), 1–11, 2021. https://doi.org/10.1007/s43762-021-00005-y.
17. C. Verdouw, B. Tekinerdogan, A. Beulens, and S. Wolfert, "Digital twins in smart farming," *Agricultural Systems*, 189, 103046, 2021. https://doi.org/10.1016/j.agsy.2020.103046.

18. A. Martínez-Gutiérrez, J. Díez-González, R. Ferrero-Guillén, P. Verde, R. Álvarez, H. Perez, "Digital twin for automatic transportation in industry 4.0," *Sensors*, 21, 3344, 2021. https://doi.org/10.3390/s21103344
19. S. R. Steinhubl, E. D. Muse, P. M. Barrett, and E. J. Topol, "Off the cuff: rebooting blood pressure treatment," *The Lancet*, 388, 749, 2016. https://doi. org/10.1016/S0140-6736(16)31348-4.
20. C. Wu, A. M. Jarrett, A. Zhou, N. Elshafeey, B. E. Adrada, R. P. Candelaria, R. M. Mohamed, M. Boge, L. Huo, J. B. White, et al. "MRI-based digital models forecast patient-specific treatment responses to neoadjuvant chemotherapy in triple-negative breast cancer," *Cancer Research*, 82, 3394–3404, 2022. https://doi.org/10.1158/0008-5472.CAN-22-1329.
21. A. De Benedictis, N. Mazzocca, A. Somma, and C. Strigaro, "Digital twins in healthcare: An architectural proposal and its application in a social distancing case study," *IEEE Journal of Biomedical and Health Informatics*, 27(10), 5143–5154. https://doi.org/10.1109/JBHI.2022.3205506
22. N. K. Chakshu, I. Sazonov, and P. Nithiarasu, "Towards enabling a cardiovascular digital twin for human systemic circulation using inverse analysis," *Biomechanics and Modeling in Mechanobiology*, 20, 449–465, 2021. https://doi.org/ 10.1007/s10237-020-01393-6
23. H. AhmedL. Devoto, "The potential of a digital twin in surgery," *Surgical Innovation*, 28(4), 509–510, 2020.
24. Z. Shu, J. Wan, and D. Zhang, "Cloud-integrated cyber–physical systems for complex industrial applications," *Mobile Networks and Applications*, 21(5), 865–878, 2016.
25. J. Hodgkinson & M. Elmouttie, "Cousins, siblings and twins: A review of the geological model's place in the digital mine," *Resources*, 9(3), 24. https://doi.org/10.3390/resources9030024.
26. Lavanya Sharma et al., "Improved patient care using robotics in the healthcare industry," in *Deep Learning in Internet of Things for Next Generation Healthcare Taylor & Francis*, CRC Press, pp. 164–189, July 2024
27. Lavanya Sharma et al., "Deep learning processes in MRI images," in *Deep Learning in Internet of Things for Next Generation Healthcare Taylor & Francis*, CRC Press, pp. 210–229, July 2024.
28. Lavanya Sharma et al., "A vision-based system in healthcare," in *Technological Prospects and Social Applications of Society* 5.0, Taylor & Francis, CRC Press, pp. 17–34, June 2023.
29. Lavanya Sharma, "Computer-aided lung cancer detection and classification of ct images using convolutional neural network," in *Computer Vision and Internet of Things: Technologies and Applications*. Taylor & Francis, CRC Press, pp. 247–262, May 2022.
30. Lavanya Sharma, "Innovation and emerging computer vision and artificial intelligence technologies in coronavirus control," in *Computer Vision and Internet of Things: Technologies and Applications*. Taylor & Francis, CRC Press, pp. 177–192, May 2022.
31. Lavanya Sharma, "Computer vision in surgical operating theatre and medical imaging," *Computer Vision and Internet of Things: Technologies and Applications*. Taylor & Francis, CRC Press, pp. 75–96, May 2022.
32. Lavanya Sharma, "Preventing Security Breach in Social Media: Threats and Prevention Techniques," in *Computer Vision and Internet of Things: Technologies and Applications,"* Taylor & Francis, CRC Press, pp. 53–62, May 2022.
33. Lavanya Sharma, and Mukesh Carpenter, "Use of robotics in real-time applications," in *Computer Vision and Internet of Things: Technologies and Applications*. Taylor & Francis, CRC Press, pp. 41–50, May 2022.

34. Katsoulakis, E., Wang, Q., Wu, H. *et al.* "Digital twins for health: a scoping review," *npj Digit. Med.* 7, 77 (2024). https://doi.org/10.1038/s41746-024-01073-0.
35. The Cleveland Clinic. Randomized Controlled Trial of Digital Twin Precision Treatment: A Novel Whole Body Digital Twin Enabled Precision Treatment for Type 2 Diabetes (TPT), 2023. https://clinicaltrials.gov/ct2/show/NCT05181449.
36. Dang, J. et al. "Predictive modeling in neurocritical care using causal artificial intelligence," *World J. Crit. Care Med,* 10, 112–119, 2021.

5 Metaverse Marketing Strategies Using IoT and 5G

Subhranil Som and Asik Rahaman Jamader

5.1 INTRODUCTION

The advent of new technologies is continuously transforming how businesses approach marketing and customer engagement. Among the most promising developments in this arena is the convergence of the Metaverse with Internet of Things (IoT) and 5G technologies. These innovations are reshaping digital landscapes, creating a virtual ecosystem where brands and consumers can interact in unprecedented ways. The Metaverse, a collective virtual shared space, represents a new frontier for immersive experiences, characterized by the blending of physical and digital worlds. Through the Metaverse, users can engage with digital environments that closely mimic the real world or represent entirely new realms, making it fertile ground for brand interactions, product placements, and marketing strategies that go beyond traditional advertising methods. The IoT plays a critical role in this transformation by linking the physical and digital worlds through a network of connected devices. IoT devices collect and transmit data in real time, providing marketers with actionable insights into consumer behavior, preferences, and habits. This information is vital for crafting highly personalized marketing strategies. In the context of the Metaverse, IoT data can be leveraged to create tailored virtual experiences for users, from personalized avatars to customized virtual environments. For instance, smart devices in a user's home could track their preferences and activities, sending this data to the Metaverse platform, where it can be used to offer personalized product suggestions or interactive virtual experiences. This level of personalization goes beyond traditional data collection methods, offering a deeper, more intimate understanding of the consumer, thus creating more engaging marketing opportunities [1].

At the same time, 5G technology provides the necessary infrastructure to support the vast amount of data generated by IoT devices and ensure seamless interaction within the Metaverse. 5G's ultra-fast speeds, low latency, and massive data capacity are pivotal in enabling real-time, high-quality interactions in virtual environments. Without 5G, the sheer volume of data required to create realistic, immersive virtual experiences would be too much for existing networks to handle. As a result, 5G empowers businesses to deliver content and experiences within the Metaverse

DOI: 10.1201/9781003582489-5

without delays, enhancing user engagement and satisfaction. One of the key advantages of incorporating IoT and 5G into Metaverse marketing is the ability to create dynamic and responsive marketing campaigns. Traditional advertising, such as banner ads or video commercials, often feels static and one-dimensional, whereas campaigns within the Metaverse can be highly interactive and evolve based on user input [2]. For example, a virtual store in the Metaverse could respond to a user's preferences by offering them personalized discounts or product recommendations in real-time, based on data collected from their IoT devices. Additionally, through augmented reality (AR) and virtual reality (VR) technologies supported by 5G, brands can create fully immersive experiences, such as allowing users to virtually try on clothes, attend virtual concerts, or explore product features in a 3D space. These experiences not only make marketing more engaging but also help build stronger emotional connections between brands and consumers.

Furthermore, the Metaverse's 24/7 nature allows for continuous customer engagement and marketing opportunities. Unlike physical stores with operating hours, virtual spaces in the Metaverse are always accessible. This constant availability, combined with IoT data, enables marketers to reach consumers at the most opportune times. For example, if a smart refrigerator notices that a user is low on certain groceries, it could trigger a marketing campaign in the Metaverse, offering a virtual coupon or a promotional deal for those products. In Figure 5.1, the real-time marketing, enabled by the integration of IoT and 5G, is far more personalized and timely than traditional marketing methods. The potential for hyper-targeted advertising within the Metaverse is immense. With IoT collecting vast amounts of data from

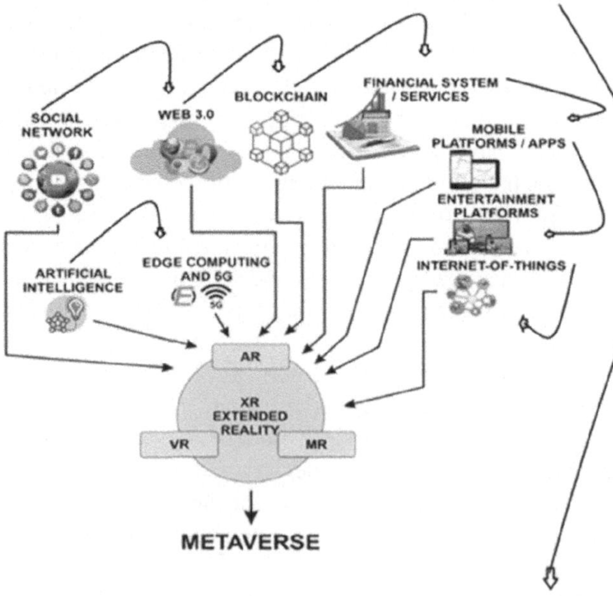

FIGURE 5.1 Metaverse marketing system

users' homes, wearables, and even vehicles, marketers can gain an unprecedented understanding of consumer preferences, habits, and needs. This allows for the creation of highly specific and personalized marketing messages, which can be delivered in real time via 5G networks. For instance, a smart home equipped with IoT devices might monitor a user's daily routine and notice they work out frequently [3]. In response, a sportswear brand could create a virtual fitness class in the Metaverse that the user can join, with personalized suggestions for workout gear based on the data collected from their IoT devices. This kind of highly relevant marketing not only grabs attention but also enhances the user experience by offering value and convenience.

In addition to enhancing personalized marketing strategies, IoT and 5G also offer marketers new opportunities to gather valuable feedback in real time. By monitoring user interactions within the Metaverse, brands can gain insights into what consumers enjoy or find useful and what needs improvement. This feedback loop allows companies to refine their marketing strategies and tailor future campaigns more effectively. Furthermore, IoT devices, such as smart speakers or wearable technology, can provide immediate feedback to marketers about how users are responding to certain products or services, enabling brands to quickly adapt to changing consumer preferences. The Metaverse, driven by IoT and 5G, also opens up new revenue streams for businesses. Virtual real estate is one example. Just as physical real estate is used for marketing and advertising in the real world, virtual spaces in the Metaverse can be rented or sold to brands. Companies can create branded environments, where users can explore products, participate in events, or engage with exclusive content. For example, a fashion brand might rent virtual space in a popular Metaverse environment to host a virtual fashion show, complete with digital versions of their latest collections that users can purchase for their avatars. With 5G supporting seamless interactions in these environments, brands can create highly engaging experiences that generate revenue while building brand loyalty.

Another exciting opportunity lies in the ability to conduct virtual product launches. Traditional product launches are often limited by geography and time, but within the Metaverse, brands can host global, 24/7 virtual events where consumers can explore new products in detail. IoT devices can enhance this experience by providing real-time updates and feedback to both the brand and the consumer. For instance, attendees of a virtual car launch might receive real-time notifications from their IoT-enabled devices about the car's specs, availability, and pricing based on their region, all while interacting with a 3D model of the car within the Metaverse. This seamless integration of virtual and physical worlds, made possible by 5G and IoT, offers an innovative approach to product promotion. Beyond consumer goods, the Metaverse also holds immense potential for industries such as real estate, travel, and hospitality. Prospective buyers can take virtual tours of homes, hotels, or travel destinations, all from the comfort of their own home, using IoT-enabled devices and 5G connectivity. IoT data can further personalize these experiences by adjusting virtual environments to match user preferences or providing real-time updates on availability and pricing [4]. For example, a potential homebuyer might take a virtual tour of a house within the Metaverse, with IoT sensors adjusting the virtual lighting to match the time of day or season. This immersive, personalized experience provides a level of engagement that is impossible to achieve through traditional marketing methods.

Despite its immense potential, the integration of IoT, 5G, and the Metaverse also presents significant challenges. Privacy and security concerns are paramount, as the extensive data collection necessary to create personalized experiences raises questions about data protection and user consent. Brands must navigate these challenges carefully to build trust with consumers. Additionally, the development and maintenance of Metaverse environments require significant investments in technology, infrastructure, and talent, which may be prohibitive for smaller businesses. Moreover, as the Metaverse continues to evolve, there will likely be ongoing debates about its regulation and governance, with implications for how brands can operate within these virtual spaces. The combination of IoT, 5G, and the Metaverse represents a transformative opportunity for marketers to create more personalized, immersive, and engaging experiences for consumers. By leveraging IoT's data collection capabilities, 5G's ultra-fast connectivity, and the Metaverse's immersive environments, brands can craft dynamic, interactive marketing campaigns that evolve based on real-time consumer input. This shift from traditional, static advertising to hyper-personalized, interactive experiences has the potential to deepen consumer engagement and foster stronger brand loyalty. However, to fully realize this potential, businesses must carefully consider the challenges associated with data privacy, security, and the substantial investments required to build and maintain a presence in the Metaverse [5]. As technology continues to advance, the intersection of IoT, 5G, and the Metaverse will undoubtedly shape the future of digital marketing in ways that are still being discovered. Brands that embrace these technologies early on and adopt innovative marketing strategies will be well-positioned to thrive in this rapidly evolving digital landscape.

5.2 REVIEW OF LITERATURE

The Metaverse, an evolving virtual space where users interact in immersive 3D environments, is emerging as a frontier that challenges traditional digital marketing models. It represents a convergence of digital technologies, including virtual and augmented reality, that creates new opportunities for brands to engage with consumers in ways that were previously unimaginable. This shift in digital landscapes has been made possible by the integration of cutting-edge technologies like the Internet of Things (IoT) and 5G, both of which play vital roles in shaping the future of marketing strategies within the Metaverse. The concept of the Metaverse extends beyond a single virtual world, encompassing multiple interconnected spaces where users can socialize, shop, work, and play. These virtual environments offer brands unprecedented access to their target audiences by allowing them to create virtual experiences that feel tangible and real. As brands move into the Metaverse, IoT and 5G technologies provide the necessary infrastructure to support these virtual spaces. IoT connects everyday physical objects to the digital world, enabling real-time data collection and analytics, while 5G ensures the fast, low-latency connectivity required for seamless interactions within virtual environments. The IoT, which refers to the network of interconnected devices that collect and exchange data, is one of the most significant technological developments influencing modern marketing. By embedding sensors in physical objects, IoT allows for the continuous collection of data on consumer behavior, preferences, and habits. For marketers, this data is invaluable, offering deep insights into how consumers interact with products and services. When applied to the Metaverse, IoT data can be used to create highly personalized experiences [6]. For example, a smart refrigerator in the real world could communicate with a virtual grocery store in the Metaverse, suggesting items based on the user's preferences and consumption patterns. This integration of physical and virtual spaces allows marketers to deliver hyper-targeted content, bridging the gap between the digital and physical worlds.

As IoT devices gather data, they provide marketers with the ability to craft individualized experiences that respond to real-time feedback. This level of personalization is already transforming how brands approach marketing in the Metaverse. Virtual environments can be tailored to reflect users' interests, preferences, and behaviors, creating a dynamic and engaging experience that fosters deeper emotional connections with brands. For instance, if a user spends time exploring virtual fitness classes in the Metaverse, IoT devices in their home gym can track their workout routines, offering personalized equipment recommendations or health tips within the virtual environment. This ability to respond to consumer behavior in real time represents a shift away from traditional, static advertising toward more fluid, interactive experiences that are more likely to resonate with consumers. In addition to personalization, IoT enables marketers to collect continuous feedback on how consumers are engaging with products and services. This feedback loop allows for more agile marketing strategies, as brands can quickly adjust their messaging, product offerings, or virtual environments based on user interactions [7]. For example, a virtual store in the Metaverse might track which products users are viewing, how long they spend interacting with certain items, and whether they proceed to make a purchase. By analyzing this data, marketers can make real-time adjustments to improve the user experience and increase conversion rates. While IoT provides the data-driven backbone of marketing in the Metaverse, 5G is the enabling technology that makes these experiences possible. 5G offers significantly faster speeds, greater bandwidth, and lower latency than previous wireless technologies, making it ideal for the data-heavy, immersive experiences of the Metaverse. Without the speed and reliability of 5G, the Metaverse's potential would be limited by slow, lagging interactions that could detract from the overall user experience. The ultra-low latency of 5G networks allows users to interact with virtual environments in real time, whether they are trying on virtual clothing, attending a virtual concert, or exploring a digital recreation of a physical store.

The speed and responsiveness of 5G also enable the creation of more complex and realistic virtual experiences. For example, virtual reality (VR) and augmented reality (AR) applications within the Metaverse demand fast data transmission to provide users with smooth, uninterrupted interactions. A seamless experience is essential for creating the sense of immersion that makes the Metaverse so appealing to consumers and marketers alike. In the context of marketing, this means that brands can create immersive, interactive experiences – such as virtual product demonstrations, 3D product visualizations, or branded virtual events without worrying about technical limitations that could hinder user engagement. Furthermore, the combination of IoT and 5G within the Metaverse enables dynamic marketing strategies that can evolve based on user input [8]. Traditional marketing methods, such as TV commercials or online banner ads, are static and offer limited interaction between the consumer and the brand. In contrast, Figure 5.2 explains Metaverse allows for highly interactive and adaptive marketing campaigns that respond to individual user behaviors. For instance, a brand might create a virtual store in the Metaverse that changes its layout or product offerings in real time based on the user's preferences, which are gathered through IoT devices. Additionally, marketers can use data collected from IoT sensors to offer users personalized promotions, discounts, or rewards within the virtual environment, further enhancing engagement.

Another significant advantage of IoT and 5G-enabled marketing in the Metaverse is the opportunity for brands to create immersive storytelling experiences. The Metaverse allows brands to move beyond traditional advertising formats and engage consumers in narratives that unfold over time and space within virtual environments.

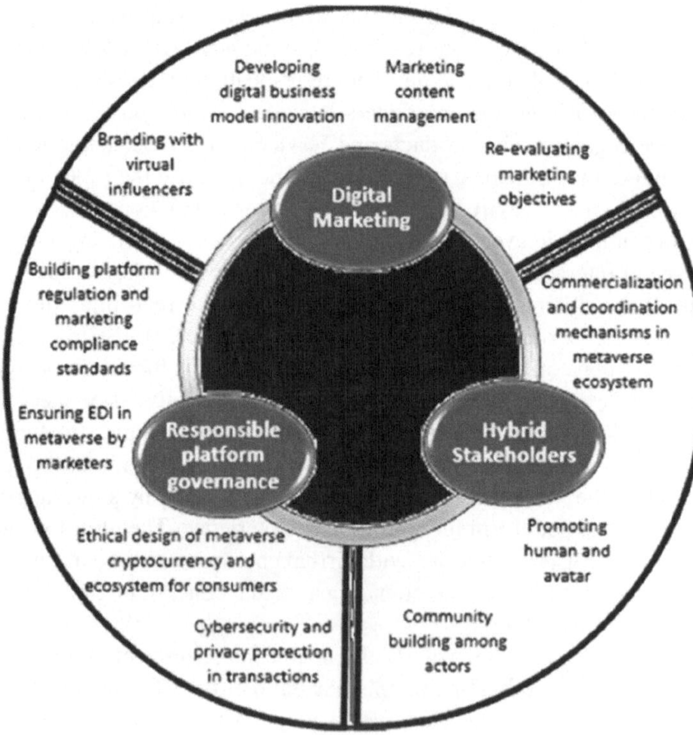

FIGURE 5.2 The road map for Metaverse marketing

For example, a fashion brand might create a virtual fashion show in the Metaverse where users can attend in real time, interact with digital versions of models, and even purchase virtual clothing for their avatars. IoT devices could further enhance this experience by syncing with the user's real-world devices, offering personalized styling tips or sending notifications when new virtual collections are available. The 24/7 nature of the Metaverse also offers unique opportunities for continuous consumer engagement. Unlike physical stores or events, virtual spaces within the Metaverse are always accessible, allowing consumers to interact with brands at any time of day or night. This level of accessibility creates a new dynamic for marketing strategies, as brands can reach consumers when they are most likely to engage, whether that is during a lunch break or late at night. IoT data can provide further insights into when and how users are engaging with the Metaverse, allowing marketers to optimize the timing and content of their campaigns.

In addition to creating personalized, immersive experiences, the integration of IoT and 5G in the Metaverse offers new opportunities for monetization. Virtual real estate within the Metaverse, for instance, is becoming a valuable commodity, as brands seek to create branded environments where they can showcase products, host events, or engage directly with consumers. Just as physical real estate is used for advertising and promotional purposes, virtual real estate in the Metaverse allows brands to create highly interactive, branded spaces that users can explore. For example, a beauty brand might create a virtual spa where users can try on virtual makeup, experiment with new hairstyles, or participate in wellness activities. With 5G providing the fast, reliable connectivity needed for these experiences, brands can create engaging virtual spaces that drive both revenue and brand loyalty.

Virtual product launches represent another area where the Metaverse, IoT, and 5G intersect to create new marketing opportunities. Traditionally, product launches are limited by geographic location and the number of attendees that can participate in an event. In the Metaverse, however, brands can host virtual product launches that are accessible to users from around the world, with no limitations on the number of attendees. These virtual launches can be highly interactive, allowing users to explore new products in detail, ask questions in real time, or participate in live demonstrations. IoT devices can further enhance the experience by providing real-time updates on product availability, pricing, or reviews based on the user's location or preferences. Beyond consumer goods, industries such as real estate, travel, and hospitality also stand to benefit from the integration of IoT, 5G, and the Metaverse. Virtual tours of real estate properties, hotels, or travel destinations can be offered within the Metaverse, allowing prospective buyers or travelers to explore these locations in a fully immersive environment. IoT devices can adjust the virtual environment to reflect real-world conditions, such as lighting or weather, providing a more realistic and personalized experience [9]. For example, a potential homebuyer could take a virtual tour of a property and see how it looks at different times of day, while IoT sensors track their preferences and adjust the tour accordingly. Despite the vast potential of marketing in the Metaverse with IoT and 5G, there are significant challenges that must be addressed. Data privacy and security concerns are at the forefront of these challenges, as the collection and use of personal data become more pervasive. The IoT ecosystem relies on continuous data collection from connected devices, which raises important questions about how this data is stored, used, and protected. Consumers may be hesitant to fully engage with the Metaverse if they are unsure how their data is being handled or if they fear their privacy could be compromised. Brands must navigate these concerns carefully, ensuring that their data practices are transparent and comply with relevant regulations.

Moreover, the cost and complexity of developing and maintaining virtual environments within the Metaverse can be prohibitive for some brands. While larger companies may have the resources to invest in cutting-edge technology, smaller businesses may struggle to keep up with the demands of creating immersive, interactive experiences. Additionally, the widespread adoption of 5G is still in progress, meaning that not all consumers will have access to the high-speed connectivity required for optimal Metaverse experiences. Brands must weigh the costs and benefits of investing in Metaverse marketing strategies, keeping in mind that the infrastructure and user base for these technologies are still evolving. As the Metaverse continues to develop, it is likely that new business models will emerge to support the growing demand for virtual experiences [10]. Brands that are early adopters of IoT and 5G in the Metaverse are likely to have a competitive advantage, as they will be able to create deeper connections with consumers through personalized, immersive interactions.

5.3 PROPOSED METHODOLOGY

This research aims to explore the integration of IoT and 5G technologies in Metaverse marketing strategies using a mixed-method approach. Both qualitative and quantitative methods will be employed to gather comprehensive data, allowing for a well-rounded understanding of how these technologies are influencing modern marketing. To begin with, case studies of companies that have successfully adopted IoT and 5G within their Metaverse marketing efforts will be analyzed. These case studies will be selected from various industries, such as retail, gaming, and entertainment, to highlight the diversity of applications. The case studies will examine specific examples of how these technologies have enhanced consumer engagement, personalization, and interactivity within the Metaverse. By studying these real-world implementations, this research can identify the strategies and practices that are most effective.

In addition to case studies, surveys will be distributed to marketing professionals and technology experts who are involved in Metaverse marketing initiatives. These surveys will include questions designed to capture both quantitative and qualitative data. Topics covered will include the level of IoT and 5G adoption, perceived benefits, challenges encountered during implementation, and their impact on brand-consumer interactions. Closed-ended questions will provide numerical data, while open-ended questions will allow respondents to elaborate on their experiences and offer deeper insights into their successes and obstacles [11].

To further enrich the findings, semi-structured interviews will be conducted with experts from the fields of IoT, 5G, and marketing. The interviewees will include professionals responsible for marketing strategy development, technologists involved in the technical implementation of IoT and 5G, and industry thought leaders who can provide a broad perspective on the current trends and future direction of these technologies in marketing. The interview questions will focus on the specific role these technologies play in enhancing the consumer experience, creating immersive marketing environments, and generating business value. The qualitative data gathered from case studies and interviews will be analyzed using thematic analysis. This process will involve identifying and coding recurring themes and patterns to understand how IoT and 5G are shaping marketing strategies in the Metaverse. The data will be systematically categorized to reveal the most significant insights about the role of technology in creating more engaging and personalized consumer interactions [12].

The quantitative data from surveys will be analyzed using descriptive and inferential statistical methods. This will allow the research to quantify the extent to which IoT and 5G have impacted various marketing metrics, such as consumer engagement, sales conversions, and brand loyalty. Statistical software will be used to run correlation and regression analyses, providing a clearer picture of the relationships between the adoption of these technologies and marketing success. To ensure the reliability and validity of the findings, the research will use multiple data sources. The mixed-method approach of combining case studies, surveys, and interviews will help triangulate the data, ensuring that the conclusions drawn are both comprehensive and robust [13]. By gathering data from different perspectives and employing rigorous analytical methods, the research will produce findings that are both reliable and applicable to a wide range of industries and contexts. Throughout the research process, ethical considerations will be upheld. All participants in the surveys and interviews will be fully informed about the purpose of the study, and their participation will be voluntary. Confidentiality and anonymity will be guaranteed, with data securely stored and used solely for research purposes. Informed consent will be obtained from all participants prior to their involvement in the study [14]. This methodology will enable a detailed exploration of how IoT and 5G technologies are transforming marketing strategies within the Metaverse. By combining qualitative and quantitative data, the research will offer a comprehensive analysis of how these technologies are reshaping the marketing landscape and provide actionable insights for brands looking to succeed in this new virtual environment [15].

5.4 OUTCOME OF THE STUDY

In the exploration of how IoT and 5G technologies are shaping marketing strategies within the Metaverse, several key findings emerged. The results demonstrate the profound impact that these technologies have on consumer engagement, personalization, and the overall effectiveness of marketing campaigns in virtual environments. The integration of IoT and 5G is not only enhancing the functionality of the Metaverse but also enabling brands to create more immersive, interactive, and data-driven experiences that resonate with consumers on a deeper level.

The first major finding from the case studies is that IoT plays a critical role in connecting the physical and digital worlds, allowing brands to gather real-time data from consumers and use that data to personalize virtual experiences. In many instances, companies that integrated IoT into their Metaverse strategies were able to significantly improve customer engagement by offering tailored experiences based on users' preferences, habits, and behaviors. For example, retail brands in the Metaverse have used IoT devices like smartwatches, fitness trackers, and connected home appliances to monitor consumer activities and recommend products or services in the virtual environment that align with their real-world needs. This kind of integration creates a seamless bridge between the digital and physical worlds, making the Metaverse experience more relevant and engaging for users. Another important aspect of IoT in Metaverse marketing is the ability to collect continuous feedback from consumers, which allows brands to adapt and optimize their strategies in real time. One case study from the fashion industry revealed how a major brand used IoT sensors embedded in virtual clothing to track how users interacted with different items in the Metaverse. By monitoring how long users spent looking at specific garments, how they customized their avatars' outfits, and whether they made a purchase, the brand was able to adjust its product offerings in real time to better meet consumer demand. This ability to react quickly to user behavior not only improved the overall user experience but also resulted in higher conversion rates and increased brand loyalty.

In addition to improving personalization and responsiveness, IoT also enhances interactivity within the Metaverse. Several case studies highlighted how IoT-enabled devices allow users to interact with virtual objects in more meaningful ways. For example, a major electronics company created a virtual store in the Metaverse where users could test out virtual versions of their products using IoT devices connected to their real-world home networks. This allowed users to see how products like smart speakers, lighting systems, and home security cameras would function in their own homes before making a purchase. The combination of IoT and Metaverse technology created a highly interactive and immersive experience that increased user engagement and made the purchasing decision easier. 5G technology, on the other hand, is shown to be the enabler of these immersive experiences by providing the fast, low-latency connectivity needed to support real-time interactions in the Metaverse. The case studies demonstrated that brands leveraging 5G technology were able to deliver smoother, more seamless experiences to their users, without the delays or technical glitches that could detract from immersion. For instance, in the gaming industry, a popular game developer used 5G to host large-scale multiplayer events in the Metaverse, where players could compete in real time, attend virtual concerts, or explore new in-game content without experiencing lag or interruptions. The ability to provide this level of connectivity was essential in keeping users engaged and ensuring that the Metaverse environment felt as real and dynamic as possible.

The surveys conducted with marketing professionals further support the importance of 5G in enhancing Metaverse marketing efforts. A majority of respondents indicated that the adoption of 5G had a direct impact on their ability to create engaging content and deliver it to users in real time. Many marketers cited the low-latency capabilities of 5G as a key factor in their decision to invest in Metaverse marketing, as it allowed them to create more complex and interactive campaigns that would not have been possible with slower, less reliable connections. Additionally, respondents noted that 5G's ability to handle large amounts of data enabled them to incorporate more sophisticated elements, such as virtual reality and augmented reality, into their marketing strategies, further enhancing the user experience. Another notable finding from the surveys is that 5G and IoT together enable brands to create marketing campaigns that are not only more personalized but also more scalable. Many respondents highlighted the fact that these technologies allow for the automation of many marketing processes, which makes it easier to reach large audiences with tailored content. For example, one respondent from the automotive industry explained how their company used IoT data to automatically generate personalized advertisements for different users based on their driving habits, vehicle preferences, and past interactions with the brand. With 5G providing the necessary bandwidth and speed, the company was able to deliver these advertisements to users in real time, creating a highly efficient and effective marketing campaign that reached millions of consumers across multiple virtual platforms.

The interviews with experts in IoT, 5G, and Metaverse marketing provided further insight into the challenges and opportunities associated with these technologies. Many experts emphasized the need for brands to carefully balance personalization with privacy concerns, as the increasing amount of data collected by IoT devices raises important ethical questions. Several interviewees noted that while consumers are often willing to share data in exchange for more personalized experiences, they are also becoming more aware of the potential risks associated with data breaches or misuse of personal information. As a result, brands need to be transparent about how they are collecting and using data, and they must implement robust security measures to protect that data. Another challenge identified by experts is the cost and complexity of implementing IoT and 5G technologies in the Metaverse. While larger companies may have the resources to invest in these technologies, smaller businesses may struggle to keep up. One interviewee from a tech startup explained that while they see the potential of the Metaverse for marketing, they have been hesitant to fully commit due to the high costs associated with developing virtual environments, integrating IoT devices, and ensuring reliable 5G connectivity. This highlights the need for more affordable and accessible solutions that can help smaller businesses take advantage of the opportunities offered by the Metaverse.

Despite these challenges, the experts interviewed were overwhelmingly positive about the future of IoT and 5G in Metaverse marketing. Many pointed out that as these technologies continue to evolve, they will become more accessible and easier to implement, allowing a wider range of brands to participate in the Metaverse. Several experts also mentioned that the growing consumer demand for immersive, interactive experiences will drive further innovation in this space, pushing brands to find new and creative ways to leverage IoT and 5G in their marketing efforts. One of the most significant opportunities identified by experts is the potential for IoT and 5G to revolutionize how brands approach storytelling in the Metaverse. By using these technologies to create more dynamic and personalized narratives, brands can engage consumers in new and exciting ways. For example, one interviewee from the entertainment industry discussed how their company is using IoT devices to track user behavior and create customized storylines for each individual in a virtual reality game. By analyzing data from IoT sensors, the game can adapt in real time to reflect the user's preferences, choices, and interactions, creating a highly personalized experience that feels unique to each player. With 5G providing the necessary connectivity, this kind of storytelling becomes scalable, allowing brands to reach large audiences with individualized content. The ability to create personalized, real-time experiences in the Metaverse also has significant implications for brand loyalty. Several case studies and interviews highlighted how brands that effectively use IoT and 5G to engage consumers in the Metaverse are able to build stronger emotional connections with their audience. For example, a major beauty brand created a virtual spa in the Metaverse where users could receive personalized skincare advice based on data collected from IoT-enabled beauty devices in their homes. This level of personalization made users feel more connected to the brand, as they felt that the brand understood their needs and preferences. As a result, the brand saw a significant increase in repeat customers and brand loyalty.

In terms of consumer behavior, the results also indicate that users are more likely to engage with brands in the Metaverse when the experiences are personalized and interactive. Surveys revealed that consumers are drawn to brands that offer them control over their virtual experiences, whether that means customizing their avatars, interacting with virtual products, or participating in branded events. The combination of IoT and 5G allows brands to deliver these interactive experiences in real time, making them more compelling and memorable for users. Additionally, many consumers reported that they are more likely to make purchases in the Metaverse when the shopping experience feels tailored to their individual needs, such as receiving personalized product recommendations based on their IoT data. The findings also suggest that IoT and 5G are creating new opportunities for monetization within the Metaverse. Several case studies revealed how brands are using virtual real estate to generate revenue by creating branded spaces where users can interact with products, attend events, or purchase virtual goods. For example, a major beverage company created a virtual bar in the Metaverse where users could socialize, participate in branded events, and purchase virtual drinks for their avatars. The company was able to monetize this experience by selling branded virtual items and offering exclusive content to users who attended their virtual events. With IoT and 5G enabling real-time interactions, these virtual spaces become valuable assets for brands looking to generate revenue in the Metaverse. Overall, the results of this research demonstrate that IoT and 5G are playing a transformative role in the evolution of marketing strategies within the Metaverse. These technologies enable brands to create personalized, immersive experiences that engage consumers in new and innovative ways, while also providing the infrastructure needed to support real-time interactions and data-driven decision-making. As the Metaverse continues to evolve, IoT and 5G will be essential in shaping the future of marketing, offering brands new opportunities to connect with consumers and drive business success.

5.5 DISCUSSION OF THE STUDY

The study presents a comprehensive analysis of how IoT and 5G technologies are transforming marketing strategies in the Metaverse. These emerging technologies are pivotal in creating immersive, interactive, and personalized experiences that engage consumers in innovative ways. The discussion revolves around the core findings, challenges, implications for businesses, and potential future trends that this study uncovers, all while keeping the complex technological landscape in mind. One of the most important themes that emerged from the research is the crucial role of IoT in connecting the physical and digital realms. The ability of IoT devices to gather real-time data from users and translate that data into actionable insights is at the core of Metaverse marketing. IoT's seamless integration allows marketers to gain valuable insights into consumer behavior, which can then be used to personalize the Metaverse experience. This level of personalization not only makes the experience more relevant and engaging for users but also fosters a stronger connection between the consumer and the brand [15, 16, 17, 18].

For instance, the use of IoT-enabled devices such as smartwatches, fitness trackers, and home appliances provides brands with an opportunity to understand consumer habits in real time. This real-time understanding creates a highly personalized marketing approach that enables brands to recommend specific products or services in the Metaverse that align with a consumer's immediate needs or preferences. The connection between real-world activities and virtual interactions is an unprecedented opportunity to reach consumers where they are, blurring the lines between online and offline experiences. However, the study also reveals that while IoT offers immense potential for enhancing personalization, it also raises critical concerns about privacy and data security. With IoT devices collecting sensitive data about consumers' habits, preferences, and even health information, there is an increasing responsibility on brands to handle this data ethically. Consumers are becoming more aware of the risks associated with data breaches and the misuse of personal information. Therefore, brands must navigate the fine line between personalization and privacy. Transparency about data collection, consent, and security measures is essential to building trust with consumers in the Metaverse. Brands that fail to prioritize privacy may find themselves facing a backlash from increasingly privacy-conscious consumers.

Another significant finding from the study is the importance of 5G technology in enabling these real-time, immersive experiences within the Metaverse. The fast, low-latency connectivity provided by 5G is critical for delivering the kind of seamless, interactive experiences that consumers expect in virtual environments. Without 5G, the real-time interactions that make the Metaverse so engaging would be significantly hindered. For example, multiplayer gaming events, live virtual concerts, and immersive shopping experiences all rely on the ability to stream vast amounts of data with minimal delays. 5G ensures that these activities can happen without disruptions, which is crucial for maintaining user engagement and immersion. The surveys and interviews conducted in the study suggest that marketers are increasingly aware of the importance of 5G in supporting their efforts to create dynamic and interactive content in the Metaverse. Many respondents noted that 5G's low-latency capabilities were a key factor in their decision to invest in Metaverse marketing. This is because 5G allows for a more robust and interactive user experience, with the ability to support technologies such as virtual reality (VR), augmented reality (AR), and other immersive tools that require high-speed data transfer and minimal latency.

The adoption of 5G also contributes to the scalability of Metaverse marketing strategies. As more users enter the Metaverse and interact with brands, the demand for high-speed, reliable connectivity will only increase. 5G's ability to handle large volumes of data makes it possible for brands to reach a larger audience without compromising the quality of the experience. This scalability is crucial for the growth of the Metaverse as a viable marketing platform, as it ensures that brands can continue to offer engaging and personalized experiences to a global audience. One of the most interesting insights from the research is how IoT and 5G together enable brands to create real-time feedback loops that allow for the continuous optimization of marketing campaigns. For example, IoT devices can collect data on how consumers are interacting with a brand's virtual environment, and that data can then be used to make real-time adjustments to the campaign. This ability to adapt on the fly is a major advantage for brands operating in the Metaverse, where user preferences and behaviors can change rapidly. By analyzing data in real time, brands can ensure that they are always providing the most relevant and engaging experience for users, which in turn leads to higher levels of consumer satisfaction and brand loyalty.

The concept of real-time marketing optimization also has implications for the way brands approach storytelling in the Metaverse. Traditional marketing campaigns are often static, with pre-determined messages and visuals that do not change once the campaign is launched. However, in the Metaverse, the use of IoT and 5G allows for a more dynamic approach to storytelling, where the narrative can evolve based on the user's interactions. This creates a more personalized and engaging experience, as users feel that they are part of the story rather than passive observers. For example, a brand might use IoT data to customize the storyline of a virtual reality game based on the user's preferences and choices, creating a unique experience for each individual user [19, 20]. This level of personalization has the potential to revolutionize the way brands connect with consumers in virtual environments. Another important discussion point from the study is the impact of IoT and 5G on brand loyalty. As brands create more personalized and immersive experiences in the Metaverse, they have the opportunity to build stronger emotional connections with consumers. This is particularly true for brands that use IoT data to understand and anticipate consumer needs, making the virtual experience feel more tailored to the individual. Several case studies highlighted how brands that effectively use IoT and 5G in their Metaverse marketing efforts are able to create a sense of belonging and community among their users, which leads to increased brand loyalty. For example, a beauty brand that offers personalized skincare advice based on data collected from IoT-enabled devices in the user's home can create a deeper connection with the consumer by addressing their unique concerns and preferences. This level of personalization not only enhances the user experience but also fosters a sense of trust and loyalty toward the brand [21, 22].

The findings also suggest that consumers are more likely to engage with brands in the Metaverse when they are given the ability to control their virtual experiences. The interactive nature of the Metaverse, combined with the real-time capabilities of IoT and 5G, allows consumers to play an active role in shaping their virtual interactions with brands. Whether it's customizing their avatars, participating in branded events, or interacting with virtual products, users are drawn to experiences that feel unique and tailored to their preferences. This level of interactivity creates a deeper level of engagement, as users feel more invested in the experience. The more control users have over their virtual environment, the more likely they are to return to that brand in the future, creating a positive feedback loop that reinforces brand loyalty.

In terms of business implications, the study reveals both opportunities and challenges for brands looking to leverage IoT and 5G in the Metaverse. On the one hand, the ability to create personalized, real-time experiences presents a significant opportunity for brands to differentiate themselves from competitors and capture the attention of consumers in a crowded marketplace. The Metaverse offers a unique platform for brands to experiment with new marketing strategies that would not be possible in traditional digital environments, such as using IoT data to create highly customized experiences or leveraging 5G to support real-time interactions between users.

On the other hand, the study also highlights the significant challenges that brands face when implementing IoT and 5G technologies in the Metaverse. The cost of developing virtual environments, integrating IoT devices, and ensuring reliable 5G connectivity can be prohibitive, particularly for smaller businesses. Several interviewees mentioned that while they see the potential of the Metaverse for marketing, they have been hesitant to fully commit due to the high costs associated with these technologies. This suggests that there is a need for more affordable and accessible solutions that can help smaller businesses take advantage of the opportunities offered by the Metaverse. In addition to cost, the complexity of implementing IoT and 5G in the Metaverse can also be a barrier for businesses. The integration of these technologies requires a high level of technical expertise, as well as the ability to manage and analyze large amounts of data. For many businesses, especially those without a dedicated IT or data analytics team, this can be a significant challenge. Brands that want to succeed in the Metaverse will need to invest in the necessary infrastructure and talent to support these complex marketing strategies. This may include partnering with technology providers or outsourcing certain aspects of the development process to third-party vendors.

Looking ahead, the study suggests that IoT and 5G will continue to play a central role in the evolution of marketing strategies within the Metaverse. As these technologies become more advanced and accessible, brands will have even greater opportunities to create immersive, interactive experiences that engage consumers in new and innovative ways. Several experts interviewed for the study predicted that the future of marketing in the Metaverse will be driven by a combination of IoT, 5G, and artificial intelligence (AI), with AI playing a key role in analyzing IoT data and automating many aspects of the marketing process. For example, AI-powered algorithms could be used to analyze data from IoT devices in real time, allowing brands to make more accurate predictions about consumer behavior and preferences. This could lead to even more personalized marketing strategies, where brands are able to anticipate consumers' needs before they are even aware of them. AI could also be used to automate the creation of personalized content, such as generating customized advertisements or virtual experiences based on individual user data [23, 24, 25]. The combination of IoT, 5G, and AI has the potential to create a truly seamless and personalized marketing experience in the Metaverse, where every interaction is tailored to the user's preferences and needs. In conclusion, the study reveals that IoT and 5G technologies are playing a transformative role in the development of marketing strategies within the Metaverse.

5.6 CONCLUSION

The study demonstrates that IoT and 5G technologies are fundamentally transforming marketing strategies within the Metaverse by enabling brands to create highly immersive, interactive, and personalized experiences. IoT serves as the bridge between the physical and digital worlds, collecting real-time data from users and allowing brands to offer tailored virtual experiences that resonate with consumers. On the other hand, 5G provides the fast, low-latency connectivity necessary for seamless interactions in the Metaverse, ensuring that users can engage in dynamic, real-time experiences without interruptions.

The integration of IoT and 5G allows for deeper personalization, enhancing consumer engagement and loyalty by providing customized experiences based on individual preferences, behaviors, and needs. This personalization leads to stronger emotional connections between brands and consumers, which translates into increased brand loyalty and higher conversion rates. By using IoT data to inform marketing strategies, brands can continuously optimize their campaigns, ensuring that they remain relevant and engaging for users. While the benefits of IoT and 5G in the Metaverse are clear, the study also highlights important challenges, particularly around privacy, data security, and the costs associated with implementing these technologies. As consumers become more aware of privacy risks, brands must navigate the fine balance between offering personalized experiences and ensuring the ethical use of data. Transparency about data collection practices and strong security measures are essential to building and maintaining consumer trust in this rapidly evolving digital landscape.

The scalability of marketing efforts in the Metaverse is also a key advantage enabled by 5G technology. With its ability to support real-time, large-scale interactions, 5G allows brands to reach a global audience while maintaining the quality of the user experience. This scalability is particularly important as the Metaverse continues to grow, offering brands new opportunities for engagement and monetization.

However, the cost and complexity of implementing IoT and 5G technologies may pose a challenge for smaller businesses, which may not have the resources to fully capitalize on these opportunities. The study suggests that as these technologies become more advanced and accessible, the barriers to entry will decrease, allowing more businesses to take advantage of the possibilities offered by the Metaverse. Looking ahead, the future of marketing in the Metaverse will likely be shaped by the continued convergence of IoT, 5G, and emerging technologies like artificial intelligence (AI). AI has the potential to further enhance the personalization of Metaverse experiences by analyzing data in real time and automating the creation of customized content. The combination of IoT, 5G, and AI could lead to even more seamless and personalized interactions in the Metaverse, where brands can anticipate and respond to consumer needs with greater accuracy and speed, IoT and 5G are pivotal to the evolution of marketing in the Metaverse, offering brands the tools to create engaging, personalized, and scalable experiences that foster consumer loyalty and drive business success. As these technologies continue to develop, they will unlock new possibilities for brands to connect with their audiences in innovative and meaningful ways, cementing the Metaverse as a central platform for the future of marketing.

5.7 FUTURE DIRECTION

The future of marketing in the Metaverse, driven by IoT and 5G technologies, is poised for continuous evolution. One key area for future development is the integration of artificial intelligence (AI) to further enhance personalization. AI will likely play a larger role in analyzing the vast amounts of data generated by IoT devices, enabling brands to deliver even more tailored experiences in real time. This could include predictive algorithms that anticipate consumer needs or automated content creation that adapts to user preferences on the fly. Another important direction is the development of more secure and transparent data practices. As consumers become increasingly concerned about privacy, brands will need to prioritize ethical data management and ensure they are compliant with emerging regulations. Developing robust systems for data protection while maintaining high levels of personalization will be a critical challenge. Additionally, the expansion of 5G networks globally will open new opportunities for brands to reach consumers in previously underserved markets, making the Metaverse more inclusive and accessible. As technology costs decrease, smaller businesses will also be able to enter the Metaverse, leading to more competition and innovation. In this rapidly evolving landscape, adaptability and a focus on consumer trust will be key for brands seeking long-term success.

REFERENCES

1. Jamader, A. R., Chowdhary, S., & Shankar Jha, S. (2023). A Road Map for Two Decades of Sustainable Tourism Development Framework. In *Resilient and Sustainable Destinations After Disaster: Challenges and Strategies* (pp. 9–18). Emerald Publishing Limited.
2. Jamader, A. R., Chowdhary, S., Jha, S. S., & Roy, B. (2023). Application of Economic Models to Green Circumstance for Management of Littoral Area: A Sustainable Tourism Arrangement. *SMART Journal of Business Management Studies, 19*(1), 70–84.
3. Jamader, A. R., Dasgupta, S., & Baibhaw, G. (2024). The Resilient Brand Management Framework as a Responsive Business Strategy in the Industry 4.0 Era. In *Exploring the Use of Metaverse in Business and Education* (pp. 113–130). IGI Global.
4. Jamader, A. R., Das, P., & Acharya, B. R. (2019, May). BcIoT: blockchain based DDoS prevention architecture for IoT. In *2019 International Conference on Intelligent Computing and Control Systems (ICCS)* (pp. 377–382). IEEE.
5. Sharma, L. (2022). Innovation and Emerging Computer Vision and Artificial Intelligence Technologies in Coronavirus Control. In *Computer Vision and Internet of Things: Technologies and Applications* (pp. 177–192). Taylor & Francis, CRC Press.

6. Sharma, L. et al. (2023). A Vision-Based System in Healthcare. In *Technological Prospects and Social Applications of Society 5.0* (pp. 17–34). Taylor & Francis, CRC Press.
7. Sharma, L. et al. (2022). An Overview of Security Issues of Internet of Things. In *Computer Vision and Internet of Things: Technologies and Applications* (pp. 29–40). Taylor & Francis, CRC Press.
8. Jamader, A. R., Dasgupta, S., & Ahmad, M. (2024). Transformation of Marketing Strategy by Metaverse in the Hospitality Industry Facing Crisis. In *Research, Innovation, and Industry Impacts of the Metaverse* (pp. 213–222). IGI Global.
9. Sharma, L., & Carpenter, M. (Eds.). (2022). *Computer Vision and Internet of Things: Technologies and Applications* (1st ed.). Chapman and Hall/CRC. https://doi.org/10.1201/9781003244165
10. Kumar, J., Arora, M., & Erkol Bayram, G. (Eds.). (2024). *Exploring the Use of Metaverse in Business and Education.* IGI Global.
11. Jamshidi, M., Yahya, S. I., Nouri, L., Hashemi-Dezaki, H., Rezaei, A., & Chaudhary, M. A. (2023). A high-efficiency diplexer for sustainable 5G-enabled IOT in metaverse transportation system and smart grids. *Symmetry, 15*(4), 821.
12. Jamshidi, M., Yahya, S. I., Nouri, L., Hashemi-Dezaki, H., Rezaei, A., & Chaudhary, M. A. (2023). A super-efficient GSM triplexer for 5G-enabled IoT in sustainable smart grid edge computing and the metaverse. *Sensors, 23*(7), 3775.
13. Bhattacharya, P., Saraswat, D., Savaliya, D., Sanghavi, S., Verma, A., Sakariya, V., ... & Manea, D. L. (2023). Towards future internet: The metaverse perspective for diverse industrial applications. *Mathematics, 11*(4), 941.
14. Gao, S. (2022, June). Research on the innovation of the Internet of Things business model under the new scenario of Metaverse. In *Proceedings of the 2022 3rd International Conference on Internet and E-Business* (pp. 44–49).
15. Li, K., Cui, Y., Li, W., Lv, T., Yuan, X., Li, S., ... & Dressler, F. (2022). When internet of things meets metaverse: Convergence of physical and cyber worlds. *IEEE Internet of Things Journal, 10*(5), 4148–4173.
16. Sharma, L., & Garg, P.K. (Eds.). (2021). *Artificial Intelligence: Technologies, Applications, and Challenges* (1st ed.). Chapman and Hall/CRC. https://doi.org/10.1201/9781003140351
17. Sharma, L. (Ed.). (2020). *Towards Smart World: Homes to Cities Using Internet of Things* (1st ed.). Chapman and Hall/CRC. https://doi.org/10.1201/9781003056751
18. Sharma, L., & Garg, P.K. (Eds.). (2019). *From Visual Surveillance to Internet of Things: Technology and Applications* (1st ed.). Chapman and Hall/CRC. https://doi.org/10.1201/9780429297922
19. A. K., G. Jha, Lavanya Sharma (2019). Challenges, Potential & Future of IOT Integrated With Block Chain. *In International Journal of Recent Technology and Engineering, 8*(2S7), 530–536.
20. Jha G., Sharma L., & Gupta S. (2021). E-health in Internet of Things (IoT) in Real-Time Scenario. In Singh P.K., Wierzchoń S.T., Tanwar S., Ganzha M., & Rodrigues J.J.P.C. (eds) *Proceedings of Second International Conference on Computing, Communications, and Cyber-Security. Lecture Notes in Networks and Systems*, vol 203. Springer, Singapore. https://doi.org/10.1007/978-981-16-0733-2_48
21. Verdugo-Cedeño, M., Jaiswal, S., Ojanen, V., Hannola, L., & Mikkola, A. (2023). Simulation-Based Digital Twins Enabling Smart Services for Machine Operations: An Industry 5.0 Approach. *International Journal of Human–Computer Interaction, 40*(20), 6327–6343. https://doi.org/10.1080/10447318.2023.2254607

22. Aivaliotis, P., Georgoulias, K., & Chryssolouris, G. (2019). The Use of Digital Twin for Predictive Maintenance in Manufacturing. *International Journal of Computer Integrated Manufacturing*, *32*(11), 1067–1080. https://doi.org/10.1080/0951192X.2019.1686173
23. Barricelli, B. R., Casiraghi, E., & Fogli, D. (2019). A Survey on Digital Twin: Definitions, Characteristics, Applications, and Design Implications. *IEEE Access*, *7*, 167653–167671. https://doi.org/10.1109/ACCESS.2019.2953499
24. Fuller, A., Fan, Z., Day, C., & Barlow, C. (2020). Digital Twin: Enabling Technologies, Challenges and Open Research. *IEEE Access*, *8*, 108952–108971. https://doi.org/10.1109/ACCESS.2020.2998358
25. Gjeldum, N., Aljinovic, A., Crnjac Zizic, M., & Mladineo, M. (2022). Collaborative Robot Task Allocation on an Assembly Line Using the Decision Support System. *International Journal of Computer Integrated Manufacturing*, *35*(4–5), 510–526. https://doi.org/10.1080/0951192X.2021.1946856

6 Geospatial Digital Twins: An Emerging Technology in GIS

Pradeep K. Garg and Martin Kappas

6.1 INTRODUCTION

A digital twin is a virtual model of a process, product, or service, representing the 3D configuration, and similar environmental conditions (Arun, 2017). It connects the real world with the digital world. It is a detailed virtual replica of a physical object or process, made to monitor the performance of a physical object, run various scenarios, and predict problems. The digital twin provides real-time status and performance monitoring from sensors and observations, which aids in decision-making processes, planning, and predictive capabilities. Digital twins may be used to perform different simulations, analyze the anticipated problems, and develop potential solutions. The main aim is to use relevant data for improvement of original physical characteristics (Canzoneri et al., 2021).

The concept of digital twin technology is not new and has been in use since the 1960s. The idea of digital twins appeared in David Gelernter's book *Mirror Worlds* in the early 1990s (Gelernter, 1993). Professor Michael Grieves introduced it first in the year 2002 to deal with the life cycle of a project, and since then the concept of digital twins has received momentum in several fields. John Vickers of NASA in 2010 introduced a new term called "digital twin." The digital twin technology now not only has been applied in a large number of industrial and engineering applications (Kamel Boulos and Zhang, 2021) but also used for the construction of buildings and bridges, urban planning, designing and manufacturing of complex products in industries, energy generation and transmission in power sector, etc.

In the recent past, the use of geospatial technology in creating 3D models has increased the applications of digital twins. Integration of several layers of data is possible with digital twins, and thus it improves the analysis (Mostak, 2021). For the creation of a digital twin model, Geographic Information System (GIS) is the key technology, as it can be used for accurate representations of complex systems. GIS-based digital twins can reflect change over time, showing historic, current, and future states. Digital twins can improve decision-making by predicting the outcomes.

A digital twin becomes "geospatial" when connected to geographic locations, allowing real-time, map-based visualizations of physical assets (Stevens, 2019).

DOI: 10.1201/9781003582489-6

Geospatial digital twins with spatial and temporal context provide location-based information on assets and infrastructure. It can cover buildings, agriculture, energy, industrial products, natural resources management, transport, etc. Geospatial digital twins can scale to meet changing needs from single facilities to large built and natural systems.

Digital twins rely on a fusion of core technologies. Modern technologies make digital twins more accessible. The IoT and sensors are essential for capturing real-time data to keep digital models updated with current conditions. The AI and ML analyze the data to identify trends and support predictive analytics (Arun, 2017). Cloud computing facilitates scalability, data storage, and system integration. GIS provides the location data to connect virtual assets to physical spaces. The 3D visualization tools can make complex data easy to interpret, enabling a detailed view of objects in virtual space. Geospatial data adds an extra layer of context, turning digital twins into multi-dimensional models capable of representing events and scenarios with higher accuracy. Together, these technologies produce a digital twin that mirrors real-world conditions. By integrating these technologies, geospatial digital twins provide location-specific insights that empower an accurate decision-making process (Grieves, 2014).

6.2 GEOSPATIAL DATA AND TECHNOLOGY

Geospatial technology is an emerging technology that uses geospatial data (e.g., satellite images, LiDAR data, GPS data, field data, etc.) in a Geographic Information System (GIS) to create maps and carry out geographic analysis, as shown in Figure 6.1. All such data include information about its precise location on the Earth's surface. This technology is being used for a large number of applications, such as environmental management, urban planning, water resource mapping, disaster

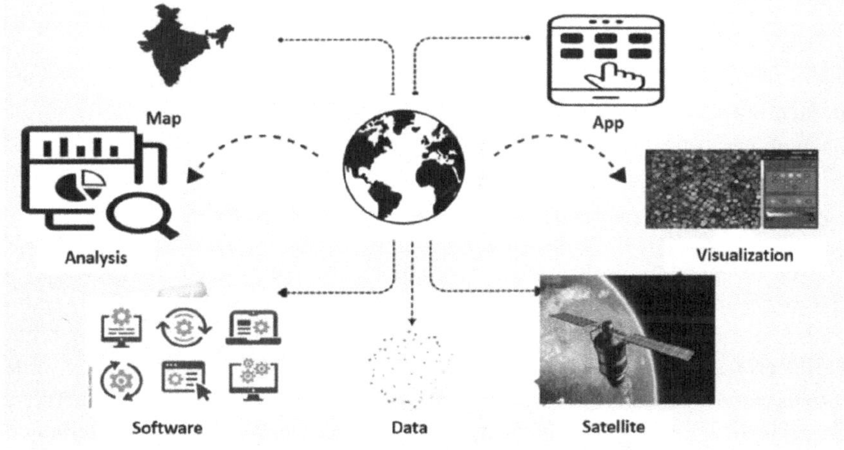

FIGURE 6.1 Various types of geospatial data used in GIS

management, health, and digital twins (Garg et al., 2022). It may also be used to generate intelligent maps and to reveal spatial patterns from a large dataset or to make a number of queries in order to derive the desired results which otherwise may not be traditionally feasible.

Geospatial data combined with longitude and latitude coordinates helps in mapping and analysis of objects/targets, and real-world phenomena. The geospatial data is available in two types: vector data and raster data (Garg, 2019). Vector data is created in the form of points, lines, and polygons that represent geographic features, such as electric poles, roads, and administrative boundaries, respectively. It is used in geospatial applications, such as networking, shortest distance, locational analysis, and navigation. In raster data, features are represented as pixels or grid cells that can be superimposed/overlaid on satellite images and scanned maps to create site suitability analysis and digital elevation models (DEMs).

High-resolution geospatial data is now frequently available which is a useful source of information to policymakers and decision-makers. Various types of geospatial data, such as remote sensing imagery and data collected from space- or airborne platforms, provide up to sub-meter spatial resolution. The GPS gives very precise locations of objects and targets facilitating the location-based analysis. The most common type of geospatial data collection tool is the use of smartphones providing location data. The use of geospatial data not only helps in developing a better understanding of the area but also supports in optimizing the resources. Today geospatial data is used for scientific planning, disaster response, environmental monitoring, infrastructure planning and development, urban planning, and supply and chain management (Foster, 2021). Government agencies are using geospatial data for health, utilities, water, weather, environment, and many others for the benefit of citizens. Industries are using it for expanding businesses, optimizing resources, and providing better services to their customers.

The use of geospatial technology is comparatively economical having a vast potential, particularly for applications where the location is crucial. It assists in making more accurate analysis and even prediction with time-series data. The technology has great potential for a wider spectrum of applications in the future due to its data resolution and accuracy. Geospatial technology can improve performance with the incorporation of artificial intelligence and smart devices, particularly in construction, precision agriculture, transportation, automobile industries, security, etc.

6.3 THE GIS TOOL

A GIS is designed to capture, store, manipulate, analyze, manage, and visualize all types of geospatial data, as shown in Figure 6.2. The objective of geospatial technology is to use geospatial data for analysis and modeling, mainly using the capabilities of GIS software. There are various data sources used in GIS, such as photogrammetric data, remote sensing images, GPS data, LiDAR data, and data from the internet (Garg, 2019). This figure shows the integration of various geospatial data techniques, including data collection, pre-processing, storage, retrieval, analysis, prediction, and visualization. The geospatial data is usually attached to the tabular data, known as

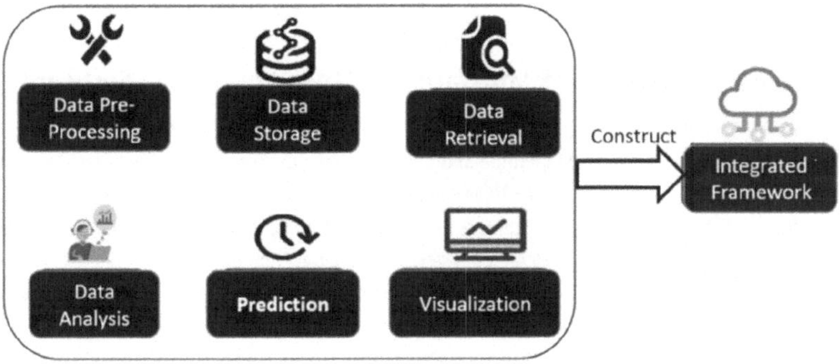

FIGURE 6.2 Various processes carried out in GIS

the attribute data. Attribute data consists of additional information about the spatial features. The integration of these two data types allows GIS to work as an effective tool for finding solutions through spatial analysis. GIS can also be used as a tool in visualization of data in 3D environment as well as in the decision-making process.

The GIS technology has grown substantially in the past 30 years or so. It has revolutionized many industries and sectors, such as infrastructure, transportation, water, environment, utilities, and health. It has the capability to transform the real world into the virtual world. In recent years, developments, like AI or machine learning algorithms, are giving ways for various innovative GIS applications. The integration of various data sources is increasing for several applications requiring detailed analysis in 3D, leading to a greater demand of GIS tools. Soon, the use of these technologies is going to become a part of our daily life work (Stevens, 2019).

Some of the cutting-edge technologies that have grown substantially in the recent past are making GIS technology more adaptable. These technologies are: mobile GIS, AI and ML, cloud computing, 3D and digital twins, automation and driverless cars, real-time and Big data analysis, and VR and AR technology. With the increase in applications, demand for real-time data and Big data for outdoor and indoor mapping, as well as facilitating various services, has increased. VR technology along with GIS is developing rapidly as it provides flexibility to users to interact with the spatial data using various innovative methods. Cloud GIS is likely to expand further due to its extensive capabilities and popularity to be used in GIS technology. Machine learning due to its capabilities to make the analysis faster and accurate is expected to be used more in the future (sharma, L 2022). The future of GIS seems to be bright as many industries and organizations sectors now understand the value of geospatial data, and are incorporating the use of GIS technology for their problem-solving. As data collection, data analysis, AR, and automated systems further develop, the demand of GIS is going to increase all around the globe. As GIS technology continues to grow, new methods of analyzing, visualizing, and managing spatial data will emerge (Mostak, 2021).

Some of the commercial GIS software and open source GIS software are given in Table 1.1. The users can select software depending on its capabilities and applications.

TABLE 1.1
Some of the Commercial GIS Software and Open Source GIS Software

S.No.	Software/Company	Commercial GIS Software
1	Autodesk	MapGuide and other products that interface with AutoCAD software
2	CADcorp	GIS software and OpenGIS standard
3	ERDAS IMAGINE	GIS, Remote Sensing, and Photogrammetry software developed by Leica Geosystems Geospatial Imaging
4	ESRI	ArcView 3.x, ArcGIS, ArcSDE, ArcIMS, and ArcWebservices
5	GRAM++	Low-cost GIS software product developed by CSRE, IIT Bombay
6	IDRISI	Developed by Clark Labs
7	Intergraph	GeoMedia, GeoMedia Profesional, GeoMedia WebMap
8	MapInfo	MapInfo Professional and MapXtreme. integrates GIS software, data, and services
9	MapPoint	Developed by Microsoft
10	My World GIS	A low-cost GIS platform for Windows and Mac OSX with robust/ intuitive geoprocessing tools, developed for educational purpose
11	SuperGeo	SuperGIS Desktop & extensions, SuperPad Suite, SuperWebGIS & extensions, SuperGIS Engine & extensions, SuperGIS Network Server and GIS services
12	SuperMap GIS	SuperMap iServer .NET/Java, SuperMap Deskpro, SuperMap Objects, SuperMap Express, SuperMap IS .NET, eSuperMap, SuperNavigation Engine, FieldMapper and services
Open Source GIS Software		
1	Chameleon	Environments for building applications with MapServer
2	GeoNetwork	A catalog application to manage spatially referenced resources
3		
4	GRASS	A complete GIS software, developed by the U.S. Army Corps of Engineers
5	GeoTools	Open source GIS toolkit written in Java, using Open Geospatial Consortium specifications
6	gvSIG	GIS software written in Java
7	ILWIS	It integrates images, vector, and thematic data
8	JUMP GIS	Java Unified Mapping Platform
9	LandSerf	Written in Java
10	MapServer	Web-based mapping server, developed by the University of Minnesota
11	MapWindow GIS	For desktop application and programming component
12	MapTools	Suite of open source GIS products and platforms
13	OpenLayers	AJAX library for accessing all kinds of geographic data layers. It is developed by MetaCarta
14	OpenStreetMap	Online map viewer, with map editing capability
15	PostGIS	Spatial extensions for the open source PostgreSQL database, allowing geospatial queries
16	SavGIS	A complete GIS software available in French, English and Spanish

(Continued)

TABLE 1.1 (CONTINUED)
Some of the Commercial GIS Software and Open Source GIS Software

S.No.	Software/Company	Commercial GIS Software
17	Quantum GIS (QGIS)	A user-friendly software that runs on Linux, Unix, Mac OSX, and Windows
18	TerraView	Desktop software that handles vector and raster data stored in a relational or geo-relational database

6.4 COMPONENTS OF A DIGITAL TWIN

The components of a digital twin include (as shown in Figure 6.3):

- **Physical Object or Process**: The real-world object or process that is represented by digital twin.
- **Digital Representation**: The 3D model that represents the real-world object or process.
- **Communication Channel**: The connection between the physical object and digital representation.
- **IoT Sensor Data**: Data collected from IoT-enabled sensors attached to the physical object.
- **Artificial Intelligence (AI)**: AI is used to accurately represent the real-world object.
- **Analytics and Simulation**: Predictive and diagnostic analytics can help identify early failures early. Simulation is a key component of digital twins.

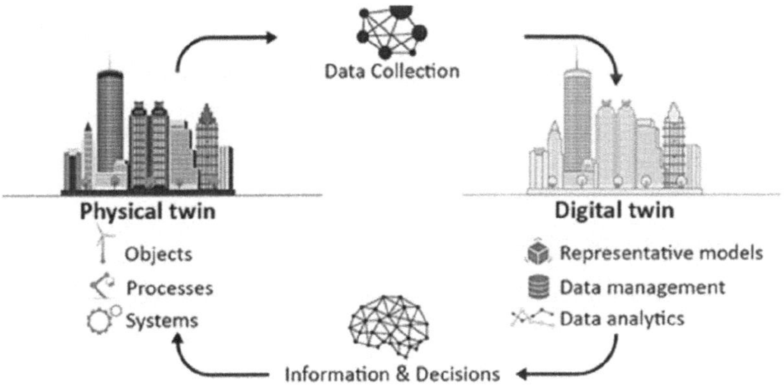

FIGURE 6.3 Components of digital twins

6.5 BENEFITS OF DIGITAL TWINS

Some of the benefits derived from digital twins are listed below.

Greater Efficiency: Digital twins can guide in making data-driven improvements, and help monitor production systems in order to achieve the highest efficiency during the manufacturing process. Organizations can improve productivity through remote assistance and monitoring. The digital twins improve operational efficiency by modeling systems with up-to-date information and testing them in dynamic conditions. With digital twins, industries can get continuous data on the performance of their products, which may lead to the innovation of new products.

Enhanced Decision-Making: Data visualizations and real-time monitoring help better decision-making.

Cost Optimization: Digital twins lower the overall costs and reduce the time to finalize the products or systems before their production. They support early detection of problems and their maintenance, which saves resources and reduces downtime. Shutdown for maintenance can be costly for a company.

Sustainability: By analyzing potential environmental impacts, organizations can prioritize eco-friendly practices, contributing to climate action and sustainability efforts.

Enhance Data Visibility: Digital twins enhance data sharing about the product or process. Past data and current data are accessible to stakeholders, encouraging broader involvement in decision-making.

Better Risk Management: Digital twins can aid in disaster planning and response, as real-time data helps companies react to issues and plan recovery strategies effectively.

Effective R&D: The digital twins allow more effective R&D of products, with huge data created about likely performance and outcomes. This can help companies make improvements in the product before starting its production.

Product Improvements: Digital twins can be used to improve the product through trials or discover some new products. We can see detailed data, find patterns, resolve quality or service issues using a digital twin, and develop better quality products.

Improve Customer Experience: Digital twins can be utilized to maximize customer satisfaction by optimizing the people, inventory, processes, and technologies of the service organization. It improves the customer experience of buying a product.

Enhance Supply Chain Agility and Resilience: A combination of emerging technologies and a digital twin can enhance end-to-end supply chain.

6.6 LIMITATIONS OF DIGITAL TWINS

Digital twins have several issues and challenges regarding data availability and quality; data integration and interoperability; data sharing, concerns about intellectual property; and data privacy and security. Some of the issues are as follows (Garg et al., 2022):

Data Management: Data cleaning is often needed for managing and updating the digital twins. The other problem is about defining the ownership of the data.

Data Security: Digital twin data passes through several networks and software, which makes its security a big challenging task. Digital twins need vast amounts of sensitive data to be collected and processed, and therefore there are chances of security breaches. Unauthorized users or hackers may have access to data and pose challenges in developing the digital twins. Challenges are more severe, especially in organizations that deal with sensitive information.

Data Cost: Digital twins being expensive and time-consuming cannot be created for each object/process, so there is a need to develop a mechanism for creating such virtual models. Huge data needs to be stored and analyzed frequently, and updating the digital twin models frequently may be a time-consuming process requiring a lot of effort.

IoT Development: Today, IoT sensors are required to collect real-time and time-series data to create digital twins of an object or process. Implementing an IoT is not straightforward as it requires a fast network infrastructure with huge storage capacity, along with data analysis and data security arrangement.

System Integration: Digital twins may require CAD software, GIS software, and other software, and they require system integration.

Complexity: The data collected of an object/process by sensors is voluminous, and may require complex data integration, algorithms, and data structuring. Any update in the design data, for example, must reflect in the latest twin model so that the customers and manufacturers get the latest information.

No Universal Standards: The non-availability of any universal standards for twin technology poses interoperability challenges. The lack of standards may also limit the exchange of data between different systems.

Addressing these challenges requires continuous innovation in GIS technologies, data management practices, and collaborative efforts among stakeholders.

6.7 EMERGING TECHNOLOGIES

A digital twin leverages modern technologies, such as smart sensors, data analytics, AI, and IoT, in order to detect and prevent system failures, improve system performance, and explore innovative techniques (Arun, 2017). Figure 6.4 illustrates the use of various modern technologies for digital twins. AI, ML, and cloud computing offer networked computational resources, and also process huge data in real time.

The AI enhances the simulation capabilities by generating various scenarios, which allows for a detailed analysis of physical assets under various conditions, and eventually aids in decision-making. The AI can speed up the process of generation of digital virtual models by learning from the past/existing data, thus minimizing the efforts needed for model development (Garg et al., 2022). The AI algorithms can easily be trained on past data to address critical issues before any failure.

The developments in sensor technologies have significantly increased the IoT-based applications in digital twins. With the convergence of AI with advanced analytics, digital twins can accurately predict the performance of an object or process,

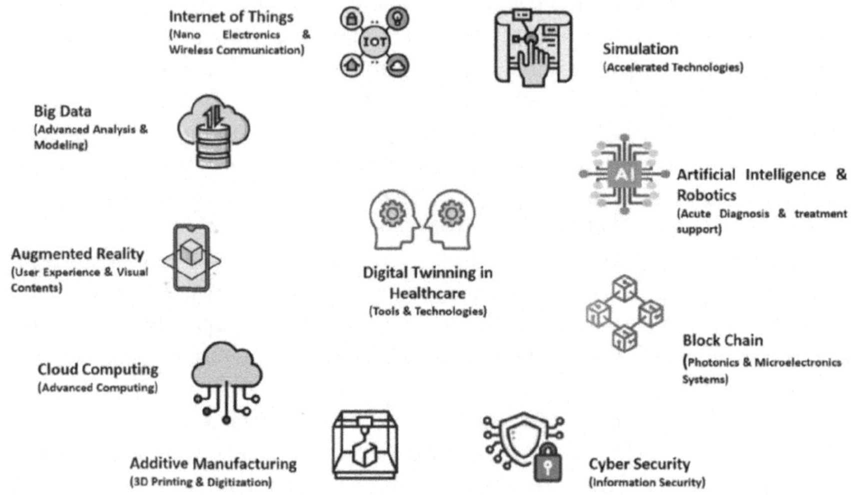

FIGURE 6.4 Various technologies used for construction of digital twins

leading to enhanced decision-making without tampering with the original, physical product. With IoT, we can get immediate feedback about an activity in progress, and apply necessary corrections immediately. With robotics, big data, and machine learning, a cost-effective twin of a system can be developed (Sharma and Garg, 2021).

Geospatial Big data, integrated with geospatial technology, can be helpful in rapid visualization, especially in a built environment. Thus, it is essential that such applications are based on the input data from the Digital Terrain Model (DTM) by using remote sensing images, GPS and BIM (Building Information Modeling) in GIS, representing the objects, such as buildings, towers, vegetation, terrain, and roads. Furthermore, levels of 3D representation result in different perspectives in terms of the physical space from a planning perspective (Park and Yang, 2020).

The GIS provides flexibility for data integration, real-time visualization, and information sharing and collaboration (Garg, 2019). GIS software is basically used for collecting, managing, analyzing, and visualizing geographical data, and allows interpretation of spatial, temporal, and geographic data to understand the relationships, patterns, and trends.

6.8 THE CONVERGENCE OF GIS AND DIGITAL TWINS

The convergence of GIS with digital twins is a technological advancement that provides the spatial dimensions essential for constructing accurate digital twins aligned with the real world. The fusion of GIS and digital twins presents a unique approach to transport planning, urban planning, and environmental management, and optimization of operational efficiencies. A digital twin becomes a geospatial digital twin when built on a foundation of geography. The addition of geospatial data turns a 3D model into a multi-dimensional model of the digital twin, with information added from different sensors (e.g., photographic, LiDAR, infrared, temperature, accelerometer, flow, etc.). A continuous flow of data from IoT devices ensures that a digital twin is matched with the changes of its real-world counterpart. The change scenarios in the digital twin allow to observe the direct and indirect impact of changes on all components of the modeled object.

Geospatial digital twins are powered by GIS, which offers crucial data for mapping digital assets to real-world locations and supports simulation of real-time events. When GIS is used in a digital twin solution, a 3D model of the object, along with its geographic mapping, component interactions, and processes, runs exactly in the same manner as in real life. GIS empowers digital twins to simulate conditions accurately, enabling organizations to model changes and study impacts beforehand. This is useful for proactive planning in industries and disaster management.

The location of geospatial data is very important as "everything happens somewhere" (Garg, 2019). The world can be represented accurately on maps using geospatial data. This information is required in several disciplines; agriculture, retail, public safety, and many more. Geospatial data involves the use of photographs and images as well as locational information of objects using photogrammetry, remote sensing, 3D scanners, Global Positioning System (GPS), GIS, and Unmanned Aerial Vehicles (UAVs)/drones (Garg, 2020). High-resolution satellite images, UAVs/drone images, and IoT-based devices and sensors provide very useful geospatial data. LiDAR data can be used to create highly accurate digital twins, while GPS can provide the location information of objects (Mostak 2021).

Massive data are needed to support the digital twins, and as the density and frequency of geospatial data are higher, the digital twin models represent the real models. More geospatial data in the digital twin system results in more innovations. For digital twinning to be effective, this data requires high-speed processing and analysis.

6.9 CONCLUSIONS

A digital twin allows for simulation, monitoring, analysis, and optimization. It has already been adopted in many industries, from product development and design to manufacturing and maintenance as well as healthcare, smart cities, aerospace and defense, and many more.

The IoT devices and sensors provide real-time data that is essential for creation of digital twins. The GIS and digital twins will be converged enhancing their capabilities, with the advancements in AI, ML, and IoT technologies. This evolution will lead to even more accurate, dynamic, and useful digital twins, opening new avenues for analysis, simulation, and decision-making. Digital twin technology is expected to grow exponentially in the near future largely due to integration of IoT, AI, VR, GIS, and Cloud computing. The global market value of digital twins was USD 12.91 billion in 2023, with North America as a major player in the global market with a share of 34.55% in 2023. It is projected to grow to USD 259.32 billion by 2032, exhibiting a CAGR of 39.8% during the forecast period (https://www.fortunebusinessinsights.com/digital-twin-market-106246).

A geospatial digital twin is constructed from inputs of many data, sensors, images, and GPS (Foster, 2021). The role of GIS in creating accurate digital twins is both foundational and transformative. By providing spatial accuracy, real-time data integration, enhanced analytical capabilities, and interoperability, GIS elevates the utility and applicability of digital twins across industries. The integration of GIS and digital twins will continue to play a pivotal role in shaping a smarter, more resilient, and more sustainable future.

REFERENCES

Arun, S. (2017). Digital Twin – An Emerging Technology for Geospatial World, Satpalda Geospatial Services. Available at: https://www.satpalda.com/blogs/digital-twin-an-emerging-technology-for-geospatial-world

Canzoneri, M., de Luca, A., Harttung, J. (2021). Digital twins: A general overview of the biopharma industry. *Adv. Biochem. Eng. Biotechnol.* 177, 167–184.

Foster, L. (2021). Authoritative Geospatial Data and Digital Twins - Denmark in 3D, techUK. Available at: https://www.techuk.org/resource/authoritive-geospatial-data-and-digital-twins-denmark-in-3d.html

Garg, P. K. (2019). *Theory and Principles of Geoinformatics*, Khanna Book Publishing, Co. Ltd., New Delhi.

Garg, P. K. (2020). *Introduction to UAV*, New Age International Pvt Ltd, Delhi.

Garg, P. K., Tripathi, N. K., Kappas, M., and Gaur, L. (Eds.) (2022). *Geospatial Data Science for Healthcare in Society 5.0*, Springer Nature, Singapore.

Gelernter, D. (1993). *Mirror Worlds: Or the Day Software Puts the Universe in A Shoebox. How It Will Happen and What It Will Mean*; Oxford University Press, Oxford, UK.

Grieves, M. (2014). Digital Twin: Manufacturing Excellence through Virtual Factory Replication (Digital Twin White Paper-2014). Available at: https://www.fortunebusinessinsights.com/digital-twin-market-106246.

Kamel Boulos, M. N., and Zhang, P. (2021). Digital Twins: From Personalised Medicine to Precision Public Health. *J. Pers. Med.*, 11, 745. https://doi.org/10.3390/jpm11080745

Mostak, T. (2021). The Future of Geospatial Analytics is in Digital Twins, Forbes. Available at: https://www.forbes.com/sites/forbestechcouncil/2021/11/18/the-future-of-geospatial-analytics-is-in-digital-twins/?sh=168ffc7dd590.

Park, J., and Yang, B. (2020). GIS-Enabled Digital Twin System for Sustainable Evaluation of Carbon Emissions: A Case Study of Jeonju City, South Korea, *Sustainability*, 12, 9186. https://doi.org/10.3390/su12219186.

Sharma, L., and Carpenter, M. (Eds.). (2022). *Computer Vision and Internet of Things: Technologies and Applications* (1st ed.). Chapman and Hall/CRC. https://doi.org/10.1201/9781003244165

Sharma, L., and Garg, P.K. (Eds.). (2021). *Artificial Intelligence: Technologies, Applications, and Challenges* (1st ed.). Chapman and Hall/CRC. https://doi.org/10.1201/9781003140351

Stevens, L. (2019). Geospatial Digital Twins. Available at: https://medium.com/@linda_29745/geospatial-digital-twins-a8e2bc886fd4

7 A Cardiac Medicine and Surgery Using Digital Twinning

A New Era in Patient Care

Lavanya Sharma, Ankit Garg, and Mukesh Carpenter

7.1 INTRODUCTION

A Digital Twins (DT) is basically a virtual replica of physical entities that allow real-time analysis and simulation. This technology is widely used in the healthcare sector. In the case of cardiovascular surgery, DT enables personalized health paths for medical practitioners in order to determine the accurate and effective intervention porotype or standards. DT are silico replicas of a person and its surrounding environment. This technology overall helps in the advancement of clinical decisions and prognosis in cardiovascular (CV) surgery. In clinical trials, this technology enables simulations of clinical situations, risk analysis, and strategies. Presently, a CV system allows to build a sophisticated mechanistic model of the human heart and vasculature by integrating data from medical imaging (MI), physiological measurement, and surgical case studies. This detailed simulation allows CV surgeons to visualize and change the anatomical structures [1–4]. This also helps in simulating several surgical scenarios and can lead to accurate decision-making.

Presently, there are various advantages of DT in CV medicine and surgery. Using this technology surgeons can get an accurate idea of procedures and complications and also modify the interventions as per the case requirement. Overall, this simulation allows CV surgeons to visualize and modify anatomical structures which leads to accurate patient outcomes [5, 6–9]. This personalized method also improves the accuracy of surgical techniques and enhances communication between the surgical team and patients, promoting a collaborative environment of case preoperative planning. The real-time applications of DT allow multimodal data into statistical models to build physiologically accurate cardiac replicas to enhance early diagnosis, workflow, and pre-op panning as shown in Figure 7.1. This technology involves the advancement of generative artificial intelligence (AI), machine learning (ML), and deep learning (DL) for dynamic and comprehensive simulation [10–13]. This

DOI: 10.1201/9781003582489-7

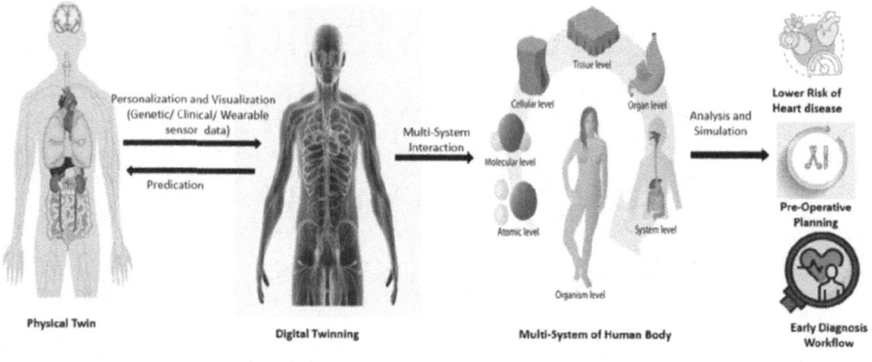

FIGURE 7.1 Digital twinning of human body shows multimodal data into statistical models to build physiologically accurate cardiac replicas to enhance early diagnosis, workflow, and pre-op panning

twinning integrates physiological, environmental, and healthcare data to build a real-time patient predication to accelerate personalized patient care. In this manner, CV digital twinning of the human body will personalize risk, prevention, and early diagnosis and treatment. A CV twinning also enables professionals to check the end result and also check the delivery decision in terms of the patient's perspective [14, 15].

7.2 HEART'S DIGITAL TWIN BY BOSTON CHILDREN'S HOSPITAL

As per the recent report, a children's hospital in Boston is presently using a heart's digital twinning (DT) to test several treatment procedures and their associated risks. The technique has resulted in lesser cases of cardiac surgeries [16–20]. Using this DT technology, cardiac surgeons can plan complex cardiac surgeries. In this manner, cardiac surgeons can now simulate procedures on "3D heart" models, which results in accurate precision and end results [2, 21–23]. Dr. David Hoganson is working as a pediatric cardiac surgeon at Boston Children's Hospital. He and his team along with the biomedical engineering team started applying digital twinning to model procedure prior to surgical procedure and as a result, improvement of patient outcomes.

In the last five years, his team has been using this cutting-edge technology associated with aerospace and automotive industries to create a roadmap for surgeries before patients are brought to the surgical room or operation theatre (OT) [24–27]. Presently, Dr. Hoganson is the Director of Computational 3D visualization program in the hospital that deals in planning and testing treatment virtually. During an event, Dr. Hoganson discussed the work and their team. His professional engineer team makes models for the admitted patients to utilize the advanced tools to perform virtual surgeries, simulation floe, and final analysis. He also discussed that some of the advanced techniques are also adopted by the other heart centers [28–31].

In collaboration with "Dassault Systemès" Dr. Hogan emphasizes more on developing "patient-centered" and "patient-specific" models and simulations to "advance at a pace they would never have dreamed would occur." According to Dr. Hogan, "The highest risk patients require scrupulous planning which is beyond the experience goes beyond experience and better planning equates better clinical outcomes" [32–36].

7.3 WORKING OF DIGITAL TWIN IN CARDIAC SURGERY

Use of alternative imaging has revolutionized the management and ultimately the outcomes of cardiac surgery and intervention in recent decades. The role of three-dimensional printing and imaging along with routine investigations is paramount in this regard. The whole process commences with the acquisition of contrast-enhanced computed tomography (CT) and magnetic resonance (MR) angiograms as shown in Figure 7.2. These imaging modalities provide excellent contrast and spatial resolution which are best suited for 3D modeling and printing which can be viewed with the naked eye or aided by virtual reality gear [1, 3]. Image acquisition is then followed by segmentation of the anatomy using specialized post-processing software and refining of design via computer-aided software [4]. The final step is physical printing of the model using high-end medical-grade printer. Two types of models are currently being used namely the solid "blood-pool" and "hollow" models [1, 3]. Solid models are created by segmenting the blood pool signal and have excellent delineation of great vessels and extracardiac anatomy including airway and esophagus; however, the visualization of intracardiac structures is limited [4]. The hollow model is generated by applying a mesh representing heart and vessel walls around the blood pool signal and removing the blood pool signal thereafter. These models have an excellent overview of the intracardiac pathology and endocardial surface imaging, enabling operating surgeons to reproduce the surgical scenes of the opened cardiac cavities beforehand [3, 4]. Each area of the heart can be represented in a

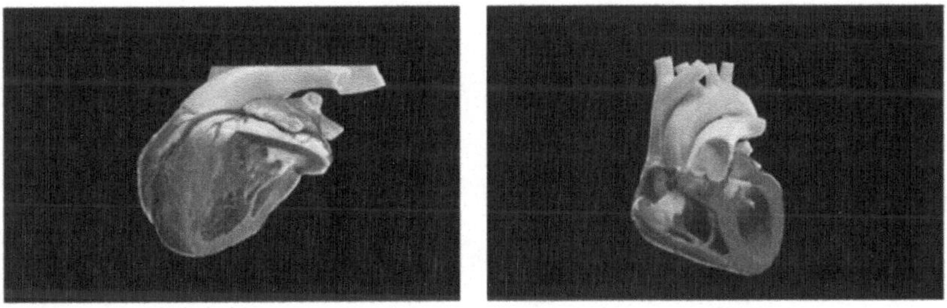

(a) (b)

FIGURE 7.2 3D images of human heart: (a) 3D model showing the right ventricle and the left ventricle in cross-section along with ventricle outflow tract commitment to the respective ventricle. (b) 3D model showing a cut section through the posterior aspect of the right ventricle showing its nearby anatomical connections

different color. Surgeons can turn any component of the model on or off and get any cross-sectional, external, or internal views of every anatomical feature. With the use of flexible models, surgeons can actually use standard surgical approaches and see the post-repair anatomy as they would for the actual case.

The major application of the 3D model is in complex congenital cardiac and difficult airway repairs as shown in Figure 7.2. The ultimate aim of most congenital cardiac surgery is to provide a biventricular repair wherein there are two separate ventricles connected to respective great vessels and the pulmonary and systemic circulation are connected in series. Assessing repair in patients with complex cardiac anatomy and those with heterotaxy using 2D imaging modality is often incomplete as the knowledge about the spatial relationship of the great vessels to ventricles and the feasibility of baffling is not possible. The 3D model helps in these situations by providing a complete spatial image with the exact nature of baffling and route of intracardiac tunneling. Presently for closure of intracardiac holes and even for complex baffling, cardiothoracic surgeons use patches the size and shape of which is a guesswork and at times may be short or long leading to a long bypass time and ultimately a guarded outcome [37–40]. Using 3D models surgeons are now able to make predetermined patches and baffles for the unique needs of the patient as per their pathology; these come in handy in the operating room as shown in Figure 7.2.

The clinical utilization of 3D models is not limited to preoperative planning rather it encompasses its use in novel areas like medical education, surgical training for young surgeons, and communication with the patient's family.

7.4 CHALLENGES OF DIGITAL TWINNING IN HEART SURGERY

The **National Science Foundation (NSF)** in collaboration with the **National Institutes of Health (NIH)** and the Food and Drug Administration (FDA) has awarded more than 6 million dollars in research funding to seven projects to explore the development of this emerging technology in health care and biomedical research.

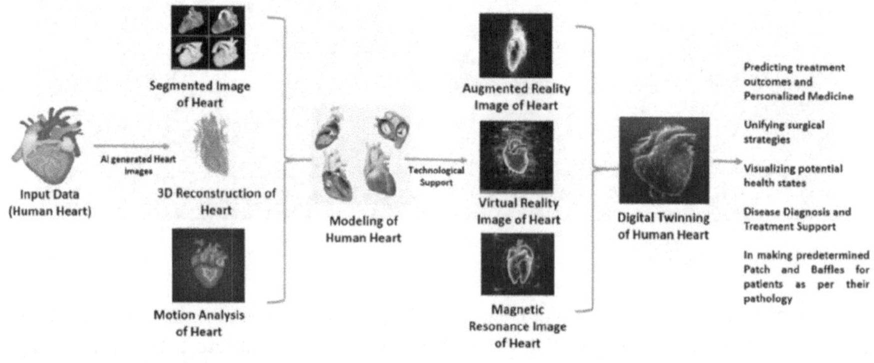

FIGURE 7.3 Working of working of digital twin in cardiac surgery and medicine

DT can be a powerful tool in the healthcare domain but has several open challenges to their implementation in heart surgery [41–45] that are listed below:

a. **Data Integration:** Combining data from various sources, such as electronic health records (EHR), sensor-based devices, and medical imaging (MI), is a very difficult task as these sources may have variations in their formats and standards.
b. **Data Accuracy:** The data that is used in digital twinning must be highly accurate and of very high quality. Inconsistency in collected data can result in inaccurate simulations.
c. **Computational Resources:** The algorithms used for simulation purposes can be resource intensive and also demand high computational power which can be costly.
d. **Model Accuracy and Reliability:** Models used in the twinning technology will be of high accuracy and also be reliable. For this purpose system needs to be timely updated with new data.
e. **Personalization and Interoperability:** DT needs to be personalized on individual parameters. **For healthcare purposes, this twinning system must have the same platform as of existing ones.**
f. **Ethical Concerns:** Patients' privacy and data security issues should be addressed effectively as twinning involves various confidential data related to patient's health.
g. **Cost and Accessibility**: The costs measured in the development and implementation of DT are very high and require expertise teams also.

7.5 FUTURE AND APPLICATIONS OF HEART TREATMENT USING DIGITAL TWINS

As per a report, researchers at the University of Maastricht and UMC Utrecht discovered a new method of pacemaker treatment for cardiac patients. By using DT, a model can be used to simulate and evaluate the future prospective of pacemakers before implanting it [46, 47]. According to Prof. Joost Lumens, a professor in the Department of Biomedical Technology, University of Maastricht, DT is a virtual replica of the human CV system and can simulate the heart's functionality and allows researchers to analyze its impact on several treatments [48]. In the last few years, DT has been widely used in the healthcare industry, mainly focusing on disease diagnosis and treatment [49–51]. By analyzing data collected from various sources or patients with pacemaker implantation, researchers were able to build DT of the patient's heart and predict the outcomes of their treatment. The use of DT in Heart surgery has various advantages. Firstly, it can assist cardiologists in pre-planning, which leads to more accurate and effective treatment for cardiac patients [52–54]. Overall, this technology results in reduction of animal testing on heart and vascular disease. Secondly, DT can be valuable in education also as it can provide students with hands-on experience and deeper knowledge of cardiac conditions and treatments [55–58]. This technology is not limited to heart disease and treatment but it has various applications in other sectors also. For example, in the case of the aviation sector, using simulation a new aircraft can be tested before being physically built. In motorsports such as Formula 1, using DT performance can be easily modified as per the requirements.

From Personalized Medicine point of view, DT makes it possible to personalized medicine for treatment in more effective manner. By accurate prediction, clinicians can make more informed decision on treatment which leads to better outcomes. Although this technology shows immense promise, it is not ready for immediate implementation. More studies and research need to be done for its effectiveness. According to Prof. Lumens it will take approximately 10 years before this technology becomes widely used in CV treatments. DT can be very useful in pacemaker implantation treatment but it can be invasive and costly. Due to the high cost of implantation every patient cannot afford and respond positively to it. The ability to check treatment outcomes beforehand can reduce the risk and treatment-associated costs [56–58].

With the present promising outcomes, further studies in this area are already in pipeline in CV medicine and surgery. The ultimate goal is to integrate this emerging technology into routine medical practice, providing patients with personalized treatment as per their past history and diagnosis. It will take a few more years to become a reality, the potential of DT on patient care and outcomes if undeniable. This innovative technology has potential to revolutionize cardiac treatment, where each and every patient receives better care based on their illness [58].

7.6 CONCLUSIONS

Digital Twins plays a vital role in medicine and cardio surgeries. In this chapter the working of DT is discussed in detail. This chapter also summarizes the open challenging issues such as data integrity, personalization and interoperability, ethics, cost, model accuracy, and reliability. In this chapter, a case of Boston's Children's Hospital where this technology is implemented is also discussed. Furthermore, the working of Digital Twin in Cardiac Surgery and clinical utilization of digital 3D model are discussed. Cardiovascular twinning enables cardiovascular surgeons and their teams to check the final result and also check the delivery decision in terms of the patient's perspective. But with present promising outcomes, further studies in this area are already in pipeline in CV medicine and surgery.

REFERENCES

1. Cardiac Surgeries. Available on: https://www.dw.com/en/using-a-hearts-digital-twin-to-reduce-cardiac-surgeries/video-68213063 [Accessed on 15 August 2024].
2. Cardiac Surgeries. Available on: https://www.washingtonpost.com/health/interactive/2024/virtual-surgery-digital-organs-doctors/ [Accessed on 10 July 2024].
3. Simulation. Available on: https://blog.3ds.com/brands/simulia/this-boston-heart-surgeon-hired-a-team-of-engineers-heres-why/ [Accessed on 12 June 2024].

4. Yoo SJ, Hussein N, Peel B, Coles J, Arsdell GS, Honjo O, Haller C, Lam CZ, Seed M, Barron D. 3D modeling and printing in congenital heart surgery: entering the stage of maturation. *Frontiers in Pediatrics* 9, 621672 (2021).
5. Anwar S, Singh GK, Miller J, Sharma M, Manning P, Billadello JJ, Eghtesady P, Woodard PK. 3D printing is a transformative technology in congenital heart disease. *JACC: Basic to Translational Science* 3(2), 294–312 (2018).
6. Sharma L, Carpenter M. (Eds.). *Computer Vision and Internet of Things: Technologies and Applications* (1st ed.). Chapman and Hall/CRC (2022). https://doi.org/10.1201/9781003244165
7. Sharma L. "Computer-Aided Lung Cancer Detection and Classification of CT Images Using Convolutional Neural Network," in *Computer Vision and Internet of Things: Technologies and Applications,* Taylor & Francis, CRC Press, pp. 247–262 (May 2022).
8. Sharma L. "Analysis of Machine Learning Techniques for Airfare Prediction," in *Computer Vision and Internet of Things: Technologies and Applications,* Taylor & Francis, CRC Press, pp. 211–231 (May 2022).
9. Cooper J, Corrias A, Gavaghan D & Noble D. Considerations for the use of cellular electrophysiology models within cardiac tissue simulations. *Progress in Biophysics and Molecular Biology* 107, 74–80 (2011).
10. Sharma L. "Innovation and Emerging Computer Vision and Artificial Intelligence Technologies in Coronavirus Control," in *Computer Vision and Internet of Things: Technologies and Applications,* Taylor & Francis, CRC Press, pp. 177–192 (May 2022).
11. Pathmanathan P., *et al.* A numerical guide to the solution of the bidomain equations of cardiac electrophysiology. *Progress in Biophysics and Molecular Biology* 102, 136–155 (2010).
12. Hunter P, Pulian A, Smail B. Modeling total heart function. *Annual Review of Biomedical Engineering* 5, 147–177 (2003).
13. Sermesant M, *et al.* Patient-specific electromechanical models of the heart for the prediction of pacing acute effects in crt: A preliminary clinical validation. *Medical Image Analysis* 16(1), 201–215 (2012).
14. Carpenter M, Carpenter D, Jangid VK, Sharma L. "Computer Vision in Surgical Operating Theatre and Medical Imaging," in *Computer Vision and Internet of Things: Technologies and Applications,* Taylor & Francis, CRC Press, pp. 75–96 (May 2022).
15. National Academies of Sciences E and M, National Academy of engineering, division on earth and life studies, division on engineering and physical sciences, board on atmospheric sciences and climate, board on life sciences, computer science and telecommunications board, Committee on Foundational Research Gaps and Future Directions for Digital Twins, Committee on Applied and Theoretical Statistics, Board on Mathematical Sciences and Analytics. *Foundational Research Gaps and Future Directions for Digital Twins.* National Academies Press, 2024.
16. Sharma L, Carpenter M (Eds.). "Use of Robotics in Real-Time Applications," *Computer Vision and Internet of Things: Technologies and Applications,* Taylor & Francis, CRC Press, pp. 41–50 (May 2022).
17. Sharma L, Sengupta S, Lohan N. "An Overview of Security Issues of Internet of Things," *Computer Vision and Internet of Things: Technologies and Applications,* Taylor & Francis, CRC Press, pp. 29–40 (May 2022).
18. Loppini A, *et al.* Competing mechanisms of stress-assisted diffusivity and stretch-activated currents in cardiac electromechanics. *Frontiers in Physiology* 9, 1714 (2018).
19. Dedè L, Gerbi A, Quarteroni A. Segregated algorithms for the numerical simulation of cardiac electromechanics in the left human ventricle. In *The Mathematics of Mechanobiology*, pp. 81–116, Springer (2020).

20. Gillette K, *et al.* Automated framework for the inclusion of a his-purkinje system in cardiac digital twins of ventricular electrophysiology. *Annals of Biomedical Engineering* 49, 3143–3153 (2021).
21. Sharma L. "The Future of Smart Cities," in *Towards Smart World: Homes to Cities using Internet of Things*, Taylor & Francis, CRC Press, pp. 1–19 (December 2020).
22. Lohan N. "Internet of things with object detection," in *Handbook of Research on Big Data and the IoT*, IGI Global, pp. 89–100 (March, 2019). https://doi.org/10.4018/978-1-5225-7432-3.ch006.
23. Corti M, *et al.* Impact of atrial fibrillation on left atrium haemodynamics: A computational fluid dynamics study. *Computers in Biology and Medicine* 150, 106143 (2022).
24. Garg PK. "Block based Adaptive Learning Rate for Moving Person Detection in Video Surveillance," in *From Visual Surveillance to Internet of Things*, Taylor & Francis, CRC Press (October 2019).
25. Garg PK. "Smart E-healthcare with Internet of Things: Current Trends Challenges, Solutions and Technologies," in *From Visual Surveillance to Internet of Things*, Taylor & Francis, CRC Press (October 2019).
26. Agarwal N. "A foresight on e-healthcare Trailblazers," in *From Visual Surveillance to Internet of Things*," Taylor & Francis, CRC Press (October 2019).
27. Viola F, Meschini V, Verzicco R. Fluid-structure-electrophysiology interaction (fsei) in the left-heart: A multi-way coupled computational model. *European Journal of Mechanics - B/Fluids* 79, 212–232 (2020).
28. Garg PK. "IoT and its applications," in *From Visual Surveillance to Internet of Things*, Taylor & Francis, CRC Press (October 2019).
29. Sharma L, Garg PK. "Future of Internet of Things," in *From Visual Surveillance to Internet of Things*, Taylor & Francis, CRC Press (October 2019).
30. Singh PK, Wierzchoń ST, Tanwar S, Ganzha M, Rodrigues JJPC (Eds.). *Proceedings of Second International Conference on Computing, Communications, and Cyber-Security. Lecture Notes in Networks and Systems*, Vol. 203. Springer, Singapore (2021). https://doi.org/10.1007/978-981-16-0733-2_48
31. Sharma S, Verma S, Kumar M. "Use of Motion Capture in 3D Animation: Motion Capture Systems, Challenges, and Recent Trends," in *2019 International Conference on Machine Learning, Big Data, Cloud and Parallel Computing (COMITCon)*, pp. 289–294 (2019). https://doi.org/10.1109/COMITCon.2019.8862448.
32. Makkar S, Sharma L. A Face Detection Using Support Vector Machine: Challenging Issues, Recent Trend, Solutions and Proposed Framework. In: Singh M, Gupta P, Tyagi V, Flusser J, Ören T, Kashyap R (Eds.), in *Advances in Computing and Data Sciences.* ICACDS 2019. Communications in Computer and Information Science, Vol. 1046. Springer, Singapore (2019). https://doi.org/10.1007/978-981-13-9942-8_1
33. Sengupta S, Kumar B. An Improved Technique for Enhancement of Satellite Images, *Journal of Physics: Conference Series* 1714, 012051 (January 2021).
34. Lohan N. Performance analysis of moving object detection using BGS techniques in visual surveillance. *International Journal of Spatiotemporal Data Science, Inderscience* 1, 22–53 (January 2019).
35. Kumar A, Jha G. "Challenges, potential & future of IOT integrated with block chain," in *International Journal of Recent Technology and Engineering* 8(2S7), 530–536 (July 2019) (indexed in Scopus). https://doi.org/ 10.35940/ijrte.B1099.0782S719.
36. Arthur R, French M, Ganguli J. *Digital twin: definition & value. American Institute of Aeronautics and Astronautics*, Tech. Rep. (2020).

37. Anand A, Jha V, Sharma S. An improved local binary patterns histograms techniques for face recognition for real time application. *International Journal of Recent Technology and Engineering* 8(2S7), 524–529 (July 2019) (indexed in Scopus). https://doi.org/10.35940/ijrte.B1098.0782S719.
38. Sharma L, Yadav DK, Singh A. "Fisher's Linear Discriminant Ratio based Threshold for Moving Human Detection in Thermal Video," in *Infrared Physics and Technology*, Elsevier (March 2016). (SCI impact factor: 1.58, Published).
39. Yadav DK. Histogram based Adaptive Learning Rate for Background Modelling and Moving Object Detection in Video Surveillance. *International Journal of Telemedicine and Clinical Practices, Inderscience* 2(1), 74–92 (June 2016). https://doi.org/10.1504/IJTMCP.2017.082107.
40. Sharma L, Carpenter M (Eds.). *Computer Vision and Internet of Things: Technologies and Applications* (1st ed.). Chapman and Hall/CRC (2022). https://doi.org/10.1201/9781003244165
41. Sharma L, Garg PK (Eds.). *Deep Learning in Internet of Things for Next Generation Healthcare* (1st ed.). Chapman and Hall/CRC (2024). https://doi.org/10.1201/9781003451846
42. Sharma L, Garg PK (Eds.). *Technological Prospects and Social Applications of Society 5.0* (1st ed.). Chapman and Hall/CRC (2023). https://doi.org/10.1201/9781003324720
43. Corral-Acero J, Margara F, Marciniak M, Rodero C, Loncaric F, Feng Y, Gilbert A, Fernandes JF, Bukhari HA, Wajdan A, et al. The 'digital twin' to enable the vision of precision cardiology. *The European Heart Journal* 41, 4556–4564 (2020).
44. Stehlik J, Schmalfuss C, Bozkurt B, Nativi-Nicolau J, Wohlfahrt P, Wegerich S, Rose K, Ray R, Schofield R, Deswal A, et al. Continuous wearable monitoring analytics predict heart failure hospitalization. *Circulation: Heart Failure* 13, e006513 (2020).
45. Singh S, Kumar B. A Machine Learning Based Predictive Model for Coronavirus Pandemic Scenario. *Journal of Physics: Conference Series* 1714, 012023 (January 2021).
46. Saraogi G, Gupta D, Rana A. Un-Supervised approach to backorder prediction using deep Autoencoder. *Recent Patents on Computer Science, Bentham* **14(8), 500–511 (2021)** (indexed in Scopus).
47. Carpenter M, Sengupta S. "Impact of IoT on Big Data Analytics and Applications in Medical Images," in *Deep Learning in Internet of Things for Next Generation Healthcare*, Taylor & Francis, CRC Press, pp. 110–134 (July 2024).
48. Sharma L, Garg P K. (Eds.). *Deep Learning in Internet of Things for Next Generation Healthcare* (1st ed.). Chapman and Hall/CRC. 2024. https://doi.org/10.1201/9781003451846
49. Carpenter M. "Improved Patient Care Using Robotics in the Healthcare Industry," in *Deep Learning in Internet of Things for Next Generation Healthcare*, Taylor & Francis, CRC Press, pp. 164–189 (July 2024).
50. Carpenter M. "Deep Learning Processes in MRI Images," in *Deep Learning in Internet of Things for Next Generation Healthcare*, Taylor & Francis, CRC Press, pp. 210–229 (July 2024).
51. Sharma L. "Foundation and Pillars of Society 5.0 and Their Roles," in *Technological Prospects and Social Applications of Society 5.0*, Taylor & Francis, CRC Press, pp. 166–198 (June 2023).
52. Hussain I, Hossain MA, Park S-J. A healthcare digital twin for diagnosis of stroke. In *Paper presented at: 2021 IEEE International Conference on Biomedical Engineering, Computer and Information Technology for Health (BECITHCON)*. IEEE, pp. 18–21 (2021).

53. Frangi AF, Rueckert D, Schnabel JA, Niessen WJ. Automatic construction of multiple-object three-dimensional statistical shape models: application to cardiac modeling. *IEEE Transactions on Medical Imaging* 21, 1151–1166 (2002).
54. Lloyd-Jones DM, Braun LT, Ndumele CE, Smith SC, Sperling LS, Virani SS, Blumenthal RS. Use of risk assessment tools to guide decision-making in the primary prevention of atherosclerotic cardiovascular disease: a special report from the American Heart Association and American College of Cardiology. *Circulation* 139, 139 (2019).
55. Digital Twin in Heart Surgeries. Available at: https://www.imperial.ac.uk/news/253154/digital-twin-heart-modelling-project-will/ [Accessed on 17 August 2024].
56. Digital Twin in Heart Surgeries. Available at: https://engineering.jhu.edu/news/natalia-trayanovas-cardiac-digital-twins-featured-in-wall-street-journal/ [Accessed on 29 July 2024].
57. Heart Surgery using Digital Twins. Available at: https://www.medicalbrief.co.za/virtual-hearts-a-game-changer-for-cardiac-surgery/ [Accessed on 29 September 2024].
58. Digital Twin in Heart Surgeries. Available at: https://medium.com/@w.gielen/the-future-of-heart-treatments-digital-twins-0666c22e3531 [Accessed on 30 September 2024].

8 Applications of Digital Twins in Personalized Medicine and Healthcare

V. Jokanović and M. Živković

8.1 INTRODUCTION

Michael Greaves is credited with creating digital twin (DT). DT is a set of virtual entities that are able to fully represent a potential or real physical object from the atomic to the macro level. In other words, a DT is a virtual presentation that serves as the digital counterpart of a physical object or process in real time [1–3].

Recently, the most attention has been paid to personal medicine as a new method for preventing and treating diseases, which includes the variation of genes, living conditions, and lifestyle of the patient. The development of big data technologies, cloud computing, and the Internet of Things (IoT) enhance the technical basis for applying a DT. This base provides detailed information about the incidence and progression of diseases enabling more accurate diagnoses and medical curing [4].

DT can be redefined as the digital replication of living or non-living physical entities, which opens the possibility of using DT in healthcare. As a concrete concept, DT is a virtual copy of human organs, tissues, cells, or micro-environment that constantly adapts to variations in online data, seeking to predict the patient's future as his real twin [5, 6]. Accordingly, DT is a very useful intelligent model that continuously self-optimizes to predict and enhance the patient's future treatments.

The technologies supporting DT are divided into data-driven statistical models and mechanical models integrating knowledge and data at multiple scales [7, 8]. Numerical models calculate construction performance, while analytical models are used for profound analysis. Besides, using artificial intelligence (AI), trained on real numerical data, enables it to derive its real-time structural performances based on data fed to it by real-time sensors. The construction of the DT model is essential for the basic DT investigations in medicine. It combines human anatomy and digital technology besides imaging and digital data collection processing, mathematical modeling, and similar diverse methods [8, 9].

The medicine used in treatment should be personalized because the body and physiology of every man are different. DT models based on the genetic code determine the cause of certain rare illnesses in the patient enabling researchers to use DT

DOI: 10.1201/9781003582489-8

technology to offer transformative therapies, shorten pharmaceutical processes, and enable precise predictions of the final treatment outcomes [8, 9].

In biology, medicine, and healthcare, the specific term biological digital twin (BDT) is introduced. BDT requires multiple modeling at all levels from subcellular to organismal levels. The main idea of a DT is to use computer simulations and models that can imitate a BT more faithfully. However, this method is still imperfect due to current limitations in understanding biological organisms [8–10].

During its analysis, DT uses four data sources: retrospective, cohort, personalized patient, and DT data. Specifically, retrospective data was generated by previous experiments and deposited in a public database. Other data are obtained from the study under investigation. The DT system has a fine structure that depends on the patient's living conditions. The advantage of DT is that it provides new sources of data to perform advanced analyses.

Accordingly, the personalized client data should be persistently measured. Finally, DT changes of the client data during the time should be updated and modified. Mathematical gene expression models use several approaches, mostly based on thermodynamic principles and Boolean models [36–38]. The DT model for gene expression monitors the dynamic activities of genes and proteins, including the effects of perturbations caused by using appropriate drugs [10, 11].

Tissue culture techniques have become available over the last decades, allowing efficient monitoring of the interaction of cells and organoids/body on chip [12]. This approach includes stem cell research, pharmacology, and toxicity testing, as integral parts of regenerative medicine. We should note that most of the protocols and procedures currently established in research laboratories are problematic. Methods of DTs could probably serve for more profound development, characterization, and production of 3D tissue culture [13, 14]. These methods are intensively researched to replace widely accepted so far methods of animal testing [15].

Accordingly, the DT method means digital replications of animate and inanimate objects, thus ensuring the exact transfer of data between natural objects and their virtual reflections. In contrast to 2D monolayer cell culture, 3D tissue culture is considered much more suitable for imitating cell behavior and organization in vivo from the aspect of the morphology and physiology of living organisms [16].

The ultimate goal of this method is to create a new tissue with a structure and functionality specific to a given organ in vivo, along with all the functional and mechanical properties of the extracellular matrix (ECM) that regulate entirely the metabolic process [17–20]. Devices for cell expansion and the creation of tissue constructs may be divided into static and dynamic cultivation systems. In static devices, the fluid and/or cellular tissue constituents do not move. Therefore, all processes in static devices are frequently determined by mass transfer, which involves supplying cells with oxygen or nutrients, growth factors, and removal of waste products. Such devices are likely not provided with online sensors for measuring temperature, pH or dissolved oxygen. These drawbacks can be overcome by using dynamic devices, in which cells/tissue constituents are subjected to fluid flow and mechanical activation. The dynamic way of culture takes place in the so-called "bioreactor," a device in which biological or biochemical processes develop in a dynamic, strictly controlled environment [21].

8.2 BACKGROUND OF DT

ML machine learning is a set of computational methods used to design an approximate mathematical function that matches the behavior of objects in the real world based on historical data. In other words, if it is given a set of data- input (X) and output (Y), then it is often possible to determine the relationship between X and Y, in the form of Y(X). If the input possesses multiple variables in the range from X0 to Xn and the output Y_0 to Yn, due to the extreme complexity, such problems are directed toward computer algorithms to "check" all these pairs of data to "determine" approximately their connection function. This approximate function generated by the algorithms is called an ML model [22].

The more input and output data make the ML model more accurate. This method is an example of supervised learning. In contrast, in clustering or unsupervised learning, the computer algorithms must discover the output (Y), in the form of patterns, clusters, or clusters with the input data (X_0, X_1, ... Xn). Therefore, this approach is best for discovering hidden patterns or relationships between data [23].

Another method of ML is reinforcement learning, within whom both the input and output (Y_0, Y_1 ... Yn) are known. In this case, the algorithms establish functions between X and Y through dynamic interaction with another test algorithm related to the environment. The environment confirms or negates the given result using the principle of trial and error causing the ML model to become more and more accurate in prediction [24, 25].

The following similar method is deep learning, in which trained algorithms have multiple hidden layers and are always applied to new data sets instead of dynamically adapting the model through continuous feedback [24, 26].

ML models are desirable because they consider factors that are not taken into account by the usual mathematical models. The complexity of bioprinting includes the entire chain of processes, from the pre-processing to the post-processing. In the pre-processing, tissue images are segmented at the level of one cell and reconstructed into a 3D tissue model with cellular resolution. In addition, ML helps predict the compatibility of different materials used in bioprinting [26, 27].

If the set of variable parameters is more than ten, it is nearly impossible to organize well-designed experiments in processing and post-processing. Additionally, ML is popular due to the multi-parameter complexity of searching for the optimal bioprinting method. Finally, MLand Big data, complex tasks of bioprinting makes to be possible [26, 27].

As already explained, DT technology contains constitutive elements such as physical objects, virtual models, their relationships, DT data, and services [28]. It is obvious from previous explanations that a physical object is a product of the functional system, which implements certain tasks and arranges those using appropriate sensors. ML and the Internet of Things (IoT) provide technical support in creating a physical object using data collection approaches, like 2D codes, data collection cards, and sensors. All of them collect data in real time and then send feedback on the data, which was the subject of processing, during the optimization process using various internet technology [29].

A virtual model is a digital model of physical objects, which is constituted from diverse variables and shows the physical properties of the route and its geometry in the virtual world. Augmented virtual (AR) reality or mixed reality (MR) technology provides support for the visualization and fusion of virtual and physical entities, while interconnections and communication between different parts of the DT model are realized using corresponding links [28, 29].

5G is a communication technology that has the characteristics of high speed and low latency, which meet the needs of the DT in mass data transmission [30]. DT data integrates various classes of information and their integration. Big data helps to extract more valuable information from massive data generated at high speed to explain the obtained result [31].

At the same time, the blockchain prevents unauthorized changes to duplicate data. The DT service model assumes a service for physical subjects and a virtual model enhancing the accuracy of the virtual model by calibrating the parameters, while AI has the ability to automatically perform data analysis and its integration with deep learning of DT data applying the method of intelligent matching of optimal algorithms, improving the exactness of data and the precision of various services [30, 31].

The main characteristic of digital twin simulation is the high accuracy of summation. High accuracy is a characteristic not only of simulation and the backbone of DT but also of simulation accuracy and real-time data collection [32]. Sensor data and numerical simulation results are essential to ensure DT accuracy, while the standardization of communication protocol and coding improves the quality of information, and modularity flexibility using specific combinations of particular models. These features of DT enable the mapping of patient data into a predictable framework that combines inductive and deductive reasoning to final result in precision medicine [31, 32].

Mechanistic models of DT can forecast the behavior of biological systems in emergency situations in the fields of health and disease because they combine the interaction of the biological components of the system. Mechanical network models usually take the form of an interconnected system of differential equations, whose solutions show how the concentration of specific molecules develops during their interaction and how they react to network inputs. Statistical models are consistent tools used to verify mechanical models using the method of inductive reasoning [34]

8.3 APPLICATION OF DT IN MEDICINE

Since the prices of IoT devices, every day, are becoming lower and lower, and simpler to implement, DT is becoming more and more present in the field of healthcare, as a new method of medical simulation for providing precise and efficient medical services. Today using artificial intelligence and other recent communication technologies, it is possible to precisely predict the cause of the client's illness and implement personalized and effective treatment [35].

In orthopedics, the technology based on human movement recording, real-time movement positions, and the 3D position of the patient's body is currently widely investigated. The lumbar position of the body can be calculated based on the data from the tracking sensor. By applying methods of inverse kinematics together with the finite element method, the DT positions of the lumbar part of the spine during different movements of the human body can be determined. Further, based on the information obtained, AI performs calculations in real time, after which these results should be compared with virtual lumbar spine through visualization. Besides, the biomechanical characteristics of the lumbar spine are also simultaneously evaluated for the purpose of monitoring and predicting its behavior. Finally, software by the development of three-dimensional (3D) virtual reality systems provides not only real-time forecasts of intradistal pressure and facet contact force but also the dynamic interaction and relationship between physical and digital space. The risks of repeated fractures should be evaluated in relation to the volume of bone with stress above the local yield strength in terms of inter-fragmentary deformations [35].

The digital models of the heart enable very efficient treatment of cardiovascular diseases. A personalized DT model was developed on the basis of personal CT images of the heart, produced before surgical intervention. This tool provides a real-time 3D positioning that allows surgeons to easily locate the disease and select the necessary equipment during surgery. Recurrent neural networks propose a methodology for inverse analysis, which DT performs with higher accuracy [35].

Using readily available blood pressure oscillations in the radial or carotid arteries as input for calculating blood pressure and using these oscillations combined with inverse analysis and by applying the deep learning methods, the development of an active DT is enabled. This DT provides registration and monitoring of the patient state, preventing further deterioration of the his health conditions. Applying non-invasive or minimally invasive measurement tools, this methodology makes it possible to avoid complex and invasive diagnostic procedures.

By creating DT models, pharmaceutical processes can be shortened, and realistic predictions of input-output biochemical reactions can be made [33]. The appliance of DT in medicine is basically oriented toward the management of chronic diseases. Recent digital technologies are directed at electroencephalogram (EEG) interpretation, intracranial pressure monitoring, and forecast simulation [36]. For example, in the case of the artificial pancreas modeling related to patients with type 1 diabetes, mathematical models that simulate insulin delivery can be harmonized with this personal DT model, which can in real time determine insulin requirements and regulate its blood insulin concentration [37, 38].

DT enables the development of an active virtual twin that continuously monitors and prevents deterioration of the patient's health conditions. In the future, it is expected that everyone will have their own DT. The accelerated development of DT will soon contribute to a much better understanding of many diseases and how to treat them most effectively.

Application of DT to the cardiovascular system includes DT of heart models and precise treatment of cardiovascular diseases. The key to DT diagnosis is a personalized model for determining pressure drop. A personalized DT model is based on computed tomography (CT) images of the heart obtained before a surgical procedure, providing real-time 3D insight into cardiac performance during surgery [38].

DT in surgery is used to create patient models for medical teams planning surgery. This method is used by many surgeons such as neurosurgeons, vascular surgeons, and interventional radiologists. A virtual model based on the application of DT is used for the development of diagnostic tools. Intracranial aneurysms can cause stroke and thrombosis, and in severe cases, instant death can be observed from the point of view of clinical, morphological, and hemodynamic parameters using CT images [38–40].

Besides, on this basis, this model can be added a part that accounts for the risk of surgical intervention in aneurysms. On the basis of 3D rotational angiography, Cure designed his 3D model of the aneurysm and surrounding vessels. Besides, the so-called Intelligent Maps model is developed to improve surgical visualization and decision-making throughout the operation by applying the latest graphics processing unit in the cloud and machine learning technology. Golse et al. have designed a mathematical model of the overall blood circulation in the body, which automatically calibrates itself and adapts to the individual characteristics of the patients. Based on these facts, it was shown that the DT model can correctly predict postoperative portal hypertension based on the speed of blood flow in the liver as an input [38–40].

8.4 DIGITAL HUMAN TWINS (DHT) AND PERSONALIZED MEDICINE

Therefore, in medicine, special attention is paid to the application of DTs, which would enable a perfect digital presentation of the health of each individual patient for the most effective prevention of the disease, more accurate diagnosis, as well as more effective treatment. Thanks to the evolution of basic technologies, the application of DTs in medicine is of great importance for the revolution of personalized medicine and its transformation in healthcare practice [41].

Digital human twins (DHT) represent a set of virtual patient replicas that include morphological (anatomical), physiological, and cognitive features, with the aim of achieving the most comprehensive digital representation of the body. A key feature of DHT is the two-way data flow between the physical entity and its virtual replica, which ensures continuous real-time updating of the virtual model and vice versa, as well as targeted interventions on the individual based on predictive simulations performed by DHT [42].

Because of the complexity of biological organisms, the construction of DHT requires the analysis and visualization of biological data at multiple levels, such as the level of genomics, transcriptomics, proteomics, and metabolomics, as well as various anatomical and biofunctional data. DHT aims to present a virtual replica of an individual or patient with his characteristic genotypic and phenotypic features, taking into account all that and the complex interaction between organisms and the environment [43].

The data is processed through advanced analytics and techniques of artificial intelligence, which are based on deep learning models, which require computing capacity capable of exceptional performance to process and analyze extensive biological data. Due to these performances, it possesses the capability of efficiently predicting interactions in the patient's body in real time. Currently, this method shows some drawbacks from the standpoint of implementing and interconnection between the various systems and systems and medical devices involved in the DHT. The adoption of standardized protocols for the representation and exchange of data DHT facilitates the communication and integration of electronic health records and medical devices [11], enabling the accurate exchange of patient-specific information [44].

The potential health benefits of DHT are enormous [16]. Disease modeling and clinical decision-making support molecular, morphological, and physiological disease, enabling not only precise prediction of the onset of symptoms but also the disease progression and complications [8]. Besides, DHTs enable the development of personalized therapies, which are highly effective, to target the specific molecular mechanism underlying the disease. Virtual replicas of organs and systems can be created using various imaging data and allow doctors to simulate the procedure, which can optimize their surgical plan before the actual operation [45].

Being digital replicas of particular patients, perfectly supporting their complex and unique interaction of physiology, anatomy, and environment, DHTs play an essential role in disease risk stratification and beyond. Using advanced AI-driven algorithms and data mining techniques, DHT manages to integrate multi-level patient information, across medical images, laboratory data, histology, etc., into complex virtual representations of an individual [45].

DHT virtual replicas enable predictive assessment of the impact of different screening approaches and interventions on the health of both the individual and the entire patient population, helping healthcare professionals to simulate them and compare their effectiveness and screening algorithms, in order to improve their accuracy [8]. By combining data from multiple sources, DHTs can analyze trends and patterns to predict the onset or progression of a disease [46].

Organs, tissues, and metabolic interactions can also be modeled to study the progression of a disease and its impact on the patient over time, as well as the patient's body response to the proposed therapy [8]. Genetic, histological, imaging, behavioral, and social data should be interconnected to form a virtual replica of the individual man and his diseases, by the algorithms linked to artificial intelligence to evaluate the effectiveness of therapy in real time.

Furthermore, DHT can implement in its software warning systems in case of serious disorders or unexpected behavior that require appropriate modification of treatment. This approach is extremely suitable for the treatment of very complex chronic diseases, such as multiple sclerosis [21]. These systems serve to understand the disease better, plan different medical procedures, and simulate the effect of the given drugs on the patient's disease [47].

Surgical planning of DHT is based on creating precise 3D models of the patient's anatomy through CT or MRI. These virtual replicas allow surgeons to visualize the patient's anatomy in detail, with more accurate localization of lesions and a deeper understanding of potential anatomical variations [2, 27]. Doctors can thus better plan and simulate surgical procedures and perform the interventions themselves. DHT provides a deeper insight into the patient's anatomy and morphologic as well as the characteristics of his lesions. These approaches provide better determination of optimal incision sites and the guidance of surgical resection in real time, with more pronounced patient safety and more efficient surgical outcomes [28]. By performing virtual experiments, it is possible to foresee potential challenges or complications of the intervention itself, before its execution [48].

DHTs can also be integrated with virtual or augmented reality technologies to provide surgeons with an even more interactive planning environment, and an even better understanding of the patient's anatomy [48].

With the improvement of numerical simulation, the need to monitor and diagnose the condition of the lumbar spine in real time using DT is increasing. During the application of DT in the field of orthopedics, not only appropriate experimental but also numerical models based on large databases are improved daily. When applying DT in the field of orthopedics, it is necessary to develop physical experimental models and numerical models based on data. DT is also used in traumas of the human lumbar spine by building an integrated DT body based on data on the bones of the lumbar spine to predict the biomechanical behavior of the particular lumbar spine at diverse positions. The lumbar position of a given human body is calculated using a virtual reality device and from the data sensors, which monitor the movements of the human body [49].

In brain tumor surgery, the development of DHT through replication of the morpho-functional organization of cortico-subcortical tissues allows neurosurgeons to precisely plan surgery, thus ensuring precise brain surgery [30]. Furthermore, DHT serves as a reference tool for the most successful visualization of the patient's anatomy in order to specify the identification of critical structures. In addition, the DHT can be increasingly enriched in real time with information from the operating room, upgrading the DHT itself [50].

The development of DHT is also important in procedures of organ transplantation because it improves the planning of the intervention itself. Based on the use of machine learning and deep learning algorithms, DHT can also predict possible complications after transplantation, such as potential organ rejection or organ failure. DHT helps in matching donors with potential recipients, comparing various factors, among which blood type and tissue compatibility are especially important. After transplantation, DHT can monitor its possible anomalies by updating the data from the transplanted organ. DHT also helps to better adapt implants or prostheses to the anatomy of a given patient based on the patient's unique anatomy [51].

Perhaps the best example of a clinically tested application of DT is an artificial pancreas model used to treat patients with type I diabetes. The artificial pancreas represents a mathematical model that simulates the patient's glucose metabolism so that in case of deviation, the pump attached to the patient injects the appropriate dose of insulin [18]. And yet so far there is no example in the literature of a biological digital twin at the subcellular (genomic) level that is based on real data [51].

Many illnesses with high morbidity and mortality result from complex pathogenesis. Yet there is evidence that patients receiving targeted therapies fared significantly better compared to those receiving generic treatments [31]. DHT provides a deeper understanding of the molecular basis of many diseases, including cancer and autoimmune diseases, showing that each patient responds differently to targeted therapy [52].

In oncology, DHT is built on the integration of multi-layered data including genomics and proteomics, extraction of radiological image features, and radiogenomics [52].

Besides, with the help of DHT, it is possible to determine the effectiveness of different drugs in different doses, through the simulation of their pharmacokinetic and pharmacodynamic interaction with a given patient, until the optimal therapy is chosen. Therefore, the development of drugs using DHT will significantly improve pharmacy not only through their clinical applications but also their development, as well as predicting the effects of new drugs or interactions between the body and such drugs, before conducting clinical trials, protecting patients from potential adverse reactions [52].

8.5 APPLICATION OF DT IN TISSUE ENGINEERING, SCAFFOLDS, AND BIOREACTORS

For more efficient cell proliferation, scalable bioreactor systems were specially constructed, which from the point of view of the specificity of the construction are divided into bioreactors with hollow fibers, bioreactors with macro-carriers, meander-type bioreactors, bioreactors in which the cultivation of snails without carriers takes place in 3D aggregates or spheroids, in alginates or other kind of hydrogels. These techniques enable the differentiation of stem cells in the desired direction. In doing so, the process of cell expansion is affected by both mechanical stress and shear stress, especially under the turbulent environment inside the bioreactor during mixing [53].

Organoid cultures seeded without scaffolding are based on the self-organization of cells on appropriate cell culture plates, the surface of which is not suitable for cell attachment, because of which the cells tend to form spheroids. Such spheroidization techniques are based on low-adhesion micro-patterned plates, on which microfluidic cell cultures levitate, or where the cells are pre-filled with magnetic nanoparticles that keep them floating by external magnetic field. These spheroids can be easily imaged by microscopic techniques, light, fluorescence, and confocal microscopy. It is understandable that such spheroids, especially multicellular spheroids, exhibit tissue-specific cell-cell contacts or cell-matrix interactions, mimicking gradients in oxygen, nutrients, waste products, or pH, and diffusion of regulatory molecules. They are extensively used for solid tumor growth modeling and metastasis studies, as well as in therapeutic procedures [54–56].

3D cell scaffold-based cultures provide its physical support favoring the secretion of the extracellular matrix with which the cells bind to the substrate, also allowing them to aggregate and migrate into the interstices within the scaffold to form a 3D, organotypic culture. Scaffolds usually have a simple structure, which favors the secretion of ECM. Depending on the specific application, they possess suitable porosity, stiffness, and mechanical stability, as well as suitable biological properties such as biocompatibility and adhesiveness. In addition, different growth factors, hormones, or other biologically active molecules are intercalated within the scaffold to improve the efficiency of cell proliferation or stem cell differentiation into a specific phenotype. Cell cultures grown on scaffolds, which are of biological or synthetic origin, are intended for use in organ chips or specific bioreactor systems. They should be biocompatible and biodegradable.

Hydroxyapatite and three calcium phosphate materials are mainly used for bone regeneration, while synthetic and natural polymers and composites are used in the production of scaffolds for other widespread purposes of tissue engineering. Scaffolds based on hydrogels, such as collagen and fibrin, as well as hybrid hydrogels are often functionalized with proteolytics and growth factors [57, 58].

Microfluidic techniques of 3D cell culture ("organ-on-a-chip" (OC) or "multi-organ-on-chip" (MOC)) use a controlled perfusion flow within the cell environment, which allows not only continuous nutrition and introduction of oxygen but also waste removal through the fluid medium [20, 56, 57]. Such techniques allow structure control in the micrometer range, which corresponds to the order of magnitude of the in vivo microenvironment, making them ideal for installing online sensors, such as sensors for temperature, pH, dissolved oxygen, impedance, and fluorescence, which enable the monitoring of the physiological state of the given cell culture.

In addition, these techniques allow mechanistic investigation of the dynamics of multi-organ interactions, or drug activity. The OC and MOC systems include investigations related to the skin, blood vessels, liver, gastrointestinal tract, blood-brain barrier, lung, kidney, and heart, among others, as well as combinations thereof [59, 60].

However, although microfluidic techniques are microfluidic techniques are well suited for drug testing, they are almost inapplicable in the creation of tissue implants. Here, a crucial role is played by the creation of the necessary microenvironment "in vivo" in a scale optimal for implantation, by creating suitable conditions for the co-culture of different types of cells, and the most effective interaction between them, with optimal tissue vascularization and its mechanical stimulation, etc. Besides, such a scaffold must be biodegradable [61, 62].

In addition, bioreactor systems should enable automated processing during routine tissue production for clinical applications, with key criteria for cell- and scaffold-based tissue constructs such as cell proliferation criteria, cell seeding on macro-porous scaffolds, and nutrient and oxygen delivery within the given tissue. It has been shown that mechanical stimulation, which refers to mechanical compression, hydrodynamic pressure, and fluid flow, is an important modulator of cell physiology, resulting in a positive impact on musculoskeletal tissue engineering, like cartilage and cardiovascular tissue formation [63].

These systems are able to support the culture of bulky 3D cell groups, but fail to imitate the microenvironment "in vivo." A major step in the development of tissue-like structures and their vascularization is the invention of decellularization techniques, which include the process of isolating the extracellular matrix (ECM) of tissue from the cells that occupied it, leaving the ECM original tissue scaffold, which then can be used in artificial regeneration of organs and tissues [64].

For the appropriate design of 3D tissue, behind a deep biological knowledge of specific tissues/organs, exceptional "engineering" knowledge, which includes not only the design of bioreactors for mass transfer and shear forces but also the design of particular materials for cell immobilization, which should be biocompatible, biodegradable, and understanding of mass transfer issues, including oxygen and various nutrient delivery, as well as removal of toxic metabolites, and cell culture and control procedures, is required. In order to control this, mathematical modeling and computational tools should be used to analyze the transport of oxygen, carbon dioxide, nutrients, and metabolites. Determining optimal engineering parameters is crucial to reduce the cost, reduce the time of experiments, and improve the efficiency and functionality of tissue constructs [65].

Modern machine design processes are carried out with the use of corresponding software, based on DT in the relevant area. The advantages of using cell culture in numerous biological researches are primarily related to the greater selectivity of in vivo tests, whereby the unnecessary sacrifice of animals is avoided, in all cases when it is not necessary, such as the most common tests of the mode and efficiency of the action of new drugs, from the point of view of their pharmacokinetics and pharmacodynamics. In addition, there are different 3D organ models for the corresponding entry point and site of action of the substance, for example, for skin, liver, pancreas, or lung [66, 67].

The main challenge related to tissue culture is the understanding of the complex behavior of cells and tissues and the engineering process, which implies the application of appropriate analytical tools and methods of predictive control, which includes the quantification and optimization of production processes of tissue engineering and the investigation of their influences on the "in vitro" environment and the quality of cultured tissue A strategy based on a DT is a knowledge-based development and engineering strategy, which implies a good knowledge of the investigated experimental system, and its metabolic understanding at different levels [68].

8.6 MATHEMATICAL MODELS USED IN DT TISSUE ENGENDERING

The main difference between various DTs is contained in the structure of the underlying algorithms and mathematical models, which are specific for the given use [69]. Accordingly, the investigation of fluid flows in 3D cell culture systems using Computational Fluid Dynamics (CFD) involves full automation of drug screening, including big data analysis, using appropriate sensor systems and software. Choice and degree of complexity of mathematics are basic parameters of their usability in DT. The model should have its own development over time, from a respectable simple structure based on newly available data and new observations to self-improve [70, 71].

Bearing in mind, that the first models are mechanistic models based on fundamental physical/chemical laws and a good understanding of the investigated system, the main advantage of these models is the high degree of possibility of extrapolation, which means if they are properly based on well-known current states, then future states from them can be predicted in advance [71].

In addition to the aforementioned models, chemical and biochemical models are also used to describe the biological functions of tissues and individual cells. Due to the extreme complexity of biological systems, which often include thousands of biological and chemical reactions, that are still in the research phase, these models are most often used to describe already established biological processes, such as, for example, cell growth based on glucose consumption. The most advanced recent models describe such systems using heuristically derived linear/nonlinear mathematical expressions. There are also so-called segregated and non-segregated models, where in non-segregated models, biological systems are considered as systems including their physiological, morphological, and genetic identities [72, 73].

Individual cell behaviors, such as different phases of the cell cycle and their interactions, can be explained on the basis of these models [74].

On the other side in compartmental models of CD, the investigated system is divided into particular parts/sections, constituted from corresponding sub-models, which include the connection of those subunits through their interactions. For this purpose, different shapes of network elements can be used, and the network is not limited to only one shape, so different combinations of them can also be used [75].

The fluid flow model CFD model uses the Navier-Stokes equation that describes fluid movement, enthalpy, heat radiation, and energy conversion. Boundary and initial conditions are then defined, involving heuristics and professional experience. Otherwise, errors may lead to questionable outcomes. Then simulations are performed and the data is examined and processed [76].

Using CFD, it is possible to investigate 3D cell expansion, by controlling the distribution of various biological and chemical constituents, such as growth factors and oxygen, pH gradient, liquid flow, shear stresses, etc. Disadvantages of the model include the ability to solve complex geometries, as well as the ability to predict future biological reactions, depending on the density of the software network and boundary conditions. CFD is extremely important in tissue engineering, in which a whole range of different phenomena are simultaneously present, including knowledge of hydrodynamics, nutrient transfer, cell growth, extracellular matrix secretion, etc. CFD is extremely important in tissue engineering, where a set of complex phenomena such as hydrodynamics, nutrient transfer, cell growth, and matrix deposition are present. Parameters such as fluid velocity, oxygen concentration gradient, and voltage, which are difficult to determine experimentally, are derived by CFD simulations and related to the kinetics of the corresponding metabolic responses [77].

The next approach is related to artificial intelligence (AI). There are mainly two branches of artificial intelligence: virtual AI and real AI. In virtual artificial intelligence, mathematical algorithms are used for data analysis and computer learning, known as machine learning. Such learning can be supervised, unsupervised, and reinforcement learning [78–82].

Supervised learning is based on learning from previous examples, where the end result is known, which is used exclusively for training a computer system. As a result, the regression confidence is determined and sent back to the training algorithm [78 –82].

Unsupervised machine learning attempts to identify hidden structures/clusters in data for cluster analysis and recognition of common patterns of patient behavior. Lopez et al. applied unsupervised machine learning to group patients based on genetic characteristics and clinical outcomes to classify and identify patients who are exposed to serious risk [83].

In reinforced machine learning, algorithms learn by finding ways to get good results, using constant positive and/or negative feedback. It is used to discover adaptive personalized therapies and to control robots in surgery [84, 85].

Hybrid Approaches: In hybrid modeling, different subclasses of models are summarized into one combined model, where the mechanistic models mechanistic models are a type of hybrid model used to describe various biological processes [86]. Effects that cannot be expressed by mechanical equations are incorporated by machine learning tools to describe complex systems in real time [86, 87].

8.7 BIOREACTORS FOR SPECIFIC CELL EXPANSION

In the therapeutic application of stem and immune cells, the percentage of growth in their number, which we denote by the proliferation rate, is extremely important. The most widely used method for the rapid proliferation of adherent cells is their seeding on micro-carriers in stirred reactors. It is important to note that the growth and differentiation of tissue cells, especially stem cells, are particularly influenced by mechanical forces of stretching, stress, compression, and shear [88].

Stem cells respond to the effects of shear, through shear-induced changes in growth behavior and differentiation. Flow-induced shear stress can be stimulatory or/and damaging to cells [89, 90].

Accelerated proliferation of stem cells and their unwanted pre-differentiation during the seeding process should be considered harmful. Only the smallest possible variation in lineage-specific differentiation is acceptable, whereby the genetic and epigenetic stability of the cells must be preserved. In order to achieve this, the cellular micro-surrounding in the bioreactor must be maintained almost unchanged as much as possible during the entire process of cell cultivation. In a complex, mostly turbulent environment, in a stirred bioreactor, the local shear rate varies, making it very difficult to relate the cellular effects of cell damage and their differentiation to the magnitude of the prevailing shear rate or shear stress [91, 92].

When it comes to liver bioengineering, in the case of the biological replacement of the liver the main request is to provide most or all detoxification and biotransformation functions specific to the liver during this process. The construction of a liver bioreactor is a special engineering challenge regarding the physicochemical requirements and scale of liver cells, as the liver is a complex structured organ built from many different cell subpopulations and a complex network of blood vessels. Typical liver constructs are obtained by "in vitro" cultivation of liver cells in/on synthetic scaffolds, which ensure cell adhesion, reorganization, proliferation, and development of its structure, due to which its biological replacement is shaped according to the microarchitecture of the liver. Most of the bioreactors used as bioartificial liver support systems are membrane bioreactors [94], which aim to prevent the cellular immune response.

Polymeric semipermeable membranes act as selective barriers that separate cells in the cell compartment, preventing immune-competent cells and proteins from passing through their wall. At the same time, they are completely permeable to nutrients and metabolites necessary for cell survival and metabolism. At the same time, they are completely permeable to nutrients and metabolites necessary for cell survival and metabolism. Engineering requests of mathematical modeling include the optimal delivery of nutrients and oxygen to cells and the successful control of concentration changes and total concentrations of biochemical signals. The application of mass transfer equations for diffuse external, trans-membrane, and internal transport allows to describe the cell metabolism, gas phase composition, membrane characteristics (particularly membrane depth), flow rate, etc. All of them affect the optimal design of a complex bioreactor system [94].

3D models of cultured skin cells in terms of permeability are exceptionally relevant in the development and exploration of pharmaceutical and cosmetic products. Therefore, considerable efforts have been made to determine the key parameters that describe their skin permeability and diffusion. Since animal experiments are under great debate, with serious differences between animal and human data, the development and commercialization of various "vitro" skin models has been encouraged. Human skin equivalents (HSE) are categorized into two main groups: epidermis-only models and full-thickness models [94, 95].

8.8 IN SILICO MODEL: ML/DT MODEL BIOPRINTING

"In silico" implies a specific computer simulation, which replaces numerous biological experiments, which is why the cost of applying this method is significantly lower than the cost of complex biological procedures, although its results still require biological validation in order to be finally accepted.

In silico experiments are mainly dependent on the improvements in structural biology because they look for the exact structural information for this kind of simulation. Accordingly, in silico digital fabrication is believed to be an essential driver of innovation and to play a major role in the new era of tissue engineering [96, 97].

As already shown when it comes to bioprinting, ML is used to optimize the printing of bioinks, which leads to an extremely significant reduction in the volume of necessary experiments with thousands of potential combinations. ML models have also been used in an in vitro study to identify basic genes within a complex gene expression profile inside different cell groups [98, 99].

Virtual experiments in the processing or post-processing of bioprinting imply an increasingly extensive application of in silico experiments with the application of ML during bioprinting. It is clear, at the same time, that ML requires a large database containing numerous clinical images of organs, their histology, immunohistochemistry, biomechanical properties, tissue and organ functions, etc. In addition, big data can be structured, unstructured, or semi-structured and is much more than traditional databases [100].

In ML, the first step is the collection of big data, which is most often diverse in the form of all kinds of diagnostic images contained in hospital databases, as well as data obtained by the internet from various world laboratories and research institutions, along with a huge scientific literature. Considering that ML has been applied successfully for cell phenotype recognition, it was reasonable to envision "in vivo 3D scanning" of living tissue models with cellular resolution. DT at the cellular level together with physical 3D bioprinting could fundamentally revolutionize biology by building tissues and organs [101].

Additionally, the application of ML in bioprinting and biofabrication has the potential to cause extremely large changes in the field of digital industry and information technologies. DT organ, which is based on very demanding mathematical models and modern software, is already imposing itself as a new source of knowledge. During the transition from an empirical to a digital approach, bioprinting will enter the digital era as a predictive technology that is increasingly based on virtual or in silico experiments [102].

Tissue engineering provides new possibilities even in tumor therapy, by imitating its microenvironment and by evaluating the clinical situation, which speaks of the seriousness of the patient's current health condition. Valles et al. combined a 3D in vitro lung cancer model based on a decellularized tissue matrix that defines a complex microenvironment for cell proliferation and an in silico model to identify the dependence of pulses emitted by cell signaling networks involved in tumor proliferation, apoptosis, and tumor propagation. A combined in vitro and in silico model enables the identification of biomarker profiles for targeted therapies [103, 104].

8.9 CONCLUSIONS

DT and DHT are used to collect patient health data, constituted from genetic information, medical history, and risk factors arising from the patient's lifestyle. All the above data are used to determine a personalized treatment plan. All this data is used to create personalized treatment plans, which are optimally adapted to the unique needs of each individual. By analyzing huge databases of a given species obtained from multiple sources, it is possible to develop personalized screening programs for early diagnosis. In addition, DT and DHT have a key role in simulating patient treatment, from a therapeutic as well as a surgical point of view. Further improvement of DHT and its future wider application will enable the momentum of personalized medicine, to unsuspected limits.

Besides, mathematical modeling and its inclusion in the DT/DHT model has great potential because it takes into account a significant number of different sources of data and knowledge that can be applied to better understand tissue engineering, as well as for studies of the action of drugs and their toxicity.

With well-planned experiments, and through essential determination of critical process parameters, mathematical models become more efficient every day, providing data that is monitored in real time. Such models are increasingly being integrated into control systems of various biological processes. They could accelerate tissue production and the process of cell cultivation with reduced costs and accelerated knowledge transfer time to the clinic. However, the realization of this vision is still very far away.

REFERENCES

1. Bauer, P., Stevens, B., and Hazeleger, W. A digital twins of Earth for the green transition. *Nat. Clim. Chang.* 2021, 11, 80–83.
2. Gettelman, A., Geer, A. J., Forbes, R. M., Carmichael, G. R., Feingold, G., Posselt, D. J., Stephens, G. L., van den Heever, S. C., Varble, A. C., and Zuidema, P. The future of Earth system prediction: Advances in model-data fusion. *Sci. Adv.* 2022, 8, eban3488.
3. Emmert-Streib, F., and Yli-Harja, O. What is a digital twin? Experimental design for a data-centric machine learning perspective in health. *Int. J. Mol. Sci.* 2022, 23, 13149.
4. Sun, T., He, X., and Li, Z. Digital twin in healthcare: Recent updates and challenges. *Digital Health* 2022, 9, 1–13.
5. Sun, T., He, X., Song, X., et al. The digital twin in medicine: a key to the future of healthcare? *Front. Med. (Lausanne)* 2022, 9, 907066.
6. Barricelli, B. R., Casiraghi, E., and Fogli, D. A survey on digital twin: definitions, characteristics, applications, and design implications. *IEEE Access* 2019, 7, 167653–167671.
7. Corral-Acero, J., Margara, F., Marciniak, M., et al. The 'digital twin' to enable the vision of precision cardiology. *Eur. Heart J.* 2020, 41, 4556–4564.
8. Peirlinck, M., Costabal, F. S., Yao, J., et al. Precision medicine in human heart modeling: perspectives, challenges, and opportunities. *Biomech. Model Mechanobiol.* 2021, 20, 803–831.
9. Tao, F., Xiao, B., Qi, Q., et al. Digital twin modeling. *J. Manuf. Syst.* 2022, 64, 372–389.
10. Croatti, A., Gabellini, M., Montagna, S., et al. On the integration of agents and digital twins in healthcare. *J. Med. Syst.* 2020, 44, 161.
11. Subramanian, K. Digital twin for drug discovery and development—the virtual liver. *J. Indian Inst. Sci.* 2020, 100, 653–662.
12. Dhaliwal, A. Three dimensional cell culture: A review. *Mater. Methods* 2012, 2, 162.
13. Geris, L. In Vivo, In Vitro, In Silico: Computational Tools for Product and Process Design in Tissue Engineering. In *Computational Modeling in Tissue Engineering*, Geris, L. (Ed.), Springer: Berlin/Heidelberg, Germany, 2013, pp. 1–15.
14. Ingber, D. E., van Mow, C., Butler, D., Niklason, L., Huard, J., Mao, J., Yannas, I., Kaplan, D., and Vunjak-Novakovic, G. Tissue engineering and developmental biology: Going biomimetic. *Tissue Eng.* 2006, 12, 3265–3283.
15. Möller, J., and Pörtner, R. Digital twins for tissue culture techniques—concepts, expectations, and state of the art. *Processes* 2021, 9, 447.
16. Sharma, L., and Garg, P.K. (Eds.). Deep Learning in Internet of Things for Next Generation Healthcare (1st ed.). Chapman and Hall/CRC, 2024. https://doi.org/10.1201/9781003451846
17. Bonnans, C., Chou, J., and Werb, Z. Remodelling the extracellular matrix in development and disease. *Nat. Rev. Mol. Cell Biol.* 2014, 15, 786–801.
18. Mouw, J. K., Ou, G., and Weaver, V. M. Extracellular matrix assembly: A multiscale deconstruction. *Nat. Rev. Mol. Cell Biol.* 2014, 15, 771–785.
19. Abbott, A. Cell culture: Biology's new dimension. *Nature* 2003, 424, 870–872.
20. Lovitt, C. J., Shelper, T. B., and Avery, V. M. Advanced cell culture techniques for cancer drug discovery. *Biology* 2014, 3, 345–367.

21. O'Brien, F. J. Biomaterials & scaffolds for tissue engineering. *Mater. Today* 2011, 14, 88–95
22. An, J., Chua, C. K., and Mironov, V. Application of machine learning in 3D bioprinting: Focus on development of big data and digital twin. *Int. J. Bioprinting* 2021, 7(1), 342.
23. Ng, W. L., Chan, A., Ong, Y. S., et al. Deep learning for fabrication and maturation of 3D bioprinted tissues and organs. *Virtual Phys. Prototyp.* 2020, 15, 340–358.
24. Meng, L., McWilliams, B., Jarosinski, W., et al. Machine learning in additive manufacturing: A Review. *JOM* 2020, 72, 1–15.
25. Goh, G., Sing, S., and Yeong, W. A review on machine learning in 3D printing: Applications, potential, and challenges. *Artif. Intell. Rev.* 2020, 54, 63–94.
26. Hamid, O. A., Eltaher, H. M., Sottile, V., et al. 3D bioprinting of a stem cell-laden, multi-material tubular composite: An approach for spinal cord repair. *Mater. Sci. Eng. C* 2020, 2020, 111707.
27. Lee, J. M., Sing, S. L., and Yeong, W. Y. Bioprinting of multimaterials with computer-aided design/computer-aided manufacturing. *Int. J. Bioprinting* 2020, 2020, 245.
28. Sun, T., He, X., Song, X., Shu, L., and Li, Z. The digital twin in medicine: A key to the future of healthcare? *Front. Med.* 2022, 9, 907066.
29. Schroeder, G. N., Steinmetz, C., Pereira, C. E., and Espindola, D. B. Digital twin data modeling with automationml and a communication methodology for data exchange. *IFAC Papers Online* 2016, 49, 12–17.
30. Zhang, P., Tao, Y.-Z., and Zhang, Z. Survey of several key technologies for 5g. *J. Commun.* 2016, 37, 15–29.
31. Chakshu, N. K., Sazonov, I., Nithiarasu, P. J. B. Towards enabling a cardiovascular digital twin for human systemic circulation using inverse analysis. *Biomech. Model. Mechanobiol.* 2021, 20, 449–465.
32. Wang, S., Lai, X., He, X., Qiu, Y., and Song, X. Building a trustworthy product-level shape-performance integrated digital twin with multifidelity surrogate model. *J. Mech. Des.* 2022, 144, 1–28.
33. Cellina, M., Cè, M., Alì, M., Irmici, G., Ibba, S., Caloro, E., Fazzini, D., Oliva, G., and Papa, S. Digital twins: The new frontier for personalized medicine? *Appl. Sci.* 2023, 13, 7940.
34. Eleftheriou, O., and Anagnostopoulos, C. Digital twins: a brief overview of applications, challenges and enabling technologies in the last decade. *Digital Twin* 2022, 2, 2.
35. Jung, A., Gsell, M. A. F., Augustin, C. M., and Plank, G. An integrated workflow for building digital twins of cardiac electromechanics-a multi-fidelity approach for personalising active mechanics. *Mathematics* 2022, 10, 823.
36. Kovatchev, B. A. Century of diabetes technology: signals, models, and artificial pancreas control. *Trends Endocrinol. Metab.* 2019, 30, 432–444.
37. Brown, S. A., Kovatchev, B. P., Raghinaru, D., Lum, J. W., Buckingham, B. A., Kudva, Y. C., et al. Six-month randomized, multicenter trial of closed-loop control in type1 diabetes. *N. Engl. J. Med.* 2019, 381, 1707.
38. Li, X., Lee, E. J., Lilja, S., Loscalzo, J., Schäfer, S., Smelik, M., et al. A dynamic single cell based framewor for digital twins to prioritize disease genes and drug targets. *Genome Med.* 2022, 14, 48.
39. Chakshu, N. K., Sazonov, I., and Nithiarasu, P. J. B. Towards enabling a cardiovascular digital twin for human systemic circulation using inverse analysis. *Biomech. Model. Mechanobiol.* 2021, 20, 449–465
40. Bruynseels, K. When nature goes digital: Routes for responsible innovation. *J. Responsible Innov.* 2020, 7, 342–360.

41. Abramovici, M., Göbel, J. C., and Savarino, P. Reconfiguration of smart products during their use phase based on virtual product twins. *CIRP Ann.* 2017, 66, 165–168.
42. Armeni, P., Polat, I., De Rossi, L. M., Diaferia, L., Meregalli, S., and Gatti, A. Digital twins in healthcare: Is it the beginning of a new era of evidence-based medicine? A Critical Review. *J. Pers. Med.* 2022, 12, 1255.
43. Kamel Boulos, M. N., and Zhang, P. Digital twins: From personalised medicine to precision public health. *J. Pers. Med.* 2021, 11, 745.
44. Vorisek, C. N., Lehne, M., Klopfenstein, S. A. I., Mayer, P. J., Bartschke, A., Haese, T., and Thun, S. Fast Healthcare Interoperability Resources (FHIR) for interoperability in health research: Systematic review. *JMIR Med. Inform.* 2022, 10, e35724.
45. Sears, V. A., and Morris, J. M. Establishing a point-of-care virtual planning and 3D printing program. *Semin. Plast. Surg.* 2022, 36, 133–148.
46. Voigt, I., Inojosa, H., Dillenseger, A., Haase, R., Akgün, K., and Ziemssen, T. Digital twins for multiple sclerosis. *Front. Immunol.* 2021, 12, 669811.
47. Armeni, P., Polat, I., De Rossi, L. M., Diaferia, L., Meregalli, S., and Gatti, A. Digital twins in healthcare: Is it the beginning of a new era of evidence-based medicine? A critical review. *J. Pers. Med.* 2022, 12, 1255.
48. Golse, N., Joly, F., Combari, P., Lewin, M., Nicolas, Q., Audebert, C., Samuel, D., Allard, M.-A., Sa Cunha, A., Castaing, D., et al. Predicting the risk of post-hepatectomy portal hypertension using a digital twin: A clinical proof of concept. *J. Hepatol.* 2021, 74, 661–669.
49. Devoto, L., Muscroft, S., and Chand, M. Highly accurate, patient specific, 3-dimensional mixed-reality model creation for surgical training and decision-making. *JAMA Surg.* 2019, 154, 968–969.
50. Cè, M., Irmici, G., Foschini, C., Danesini, G. M., Falsitta, L. V., Serio, M. L., Fontana, A., Martinenghi, C., Oliva, G., and Cellina, M. Artificial intelligence in brain tumor imaging: A step toward personalized medicine. *Curr. Oncol.* 2023, 30, 2673–2701.
51. Kovatchev, B. The year of transition from research to clinical practice. *Nat. Rev. Endocrinol.* 2018, 14, 74–76
52. Kamel Boulos, M. N., and Zhang, P. Digital twins: from personalised medicine to precision public health. *J. Pers. Med.* 2021, 11, 745.
53. Bröker, K., Sinelnikov, E., Gustavus, D., Schumacher, U., Pörtner, R., Hoffmeister, H., Lüth, S., and Dammermann, W. Mass production of highly active nk cells for cancer immunotherapy in a GMP conform perfusion bioreactor. *Front. Bioeng. Biotechnol.* 2019, 7, 194.
54. Haycock, J. W. *3D Cell Culture: A Review of Current Approaches and Techniques*, Humana Press: Totowa, NJ, USA, 2010.
55. Kunz-Schughart, L. A., Freyer, J. P., Hofstaedter, F., and Ebner, R. The use of 3-D cultures for high-throughput screening: The multicellular spheroid model. *J. Biomol. Screen.* 2004, 9, 273–285.
56. Joseph, J. S., Malindisa, S. T., and Ntwasa, M. Two-Dimensional (2D) and Three-Dimensional (3D) Cell Culturing in Drug Discovery. In *Cell Culture*, Mehanna, R. A. (Ed.), IntechOpen: London, UK, 2019.
57. Zhao, P., Gu, H., Mi, H., Rao, C., Fu, J., and Turng, L.-S. Fabrication of scaffolds in tissue engineering: A review. *Front. Mech. Eng.* 2017, 13, 107–119
58. Pina, S., Ribeiro, V. P., Paiva, O. C., Correlo, V. M., Oliveira, J. M., and Reis, R. L. Tissue engineering scaffolds. In *Handbook of Tissue Engineering Scaffolds*, Elsevier: Amsterdam, The Netherlands, 2019, Vol. 1, pp. 165–185.
59. Bhatia, S. N., and Ingber, D. E. Microfluidic organs-on-chips. *Nat. Biotechnol.* 2014, 32, 760–772.

60. Wu, Q., Liu, J., Wang, X., Feng, L., Wu, J., Zhu, X., Wen, W., and Gong, X. Organ-on-a-chip: Recent breakthroughs and future prospects. *Biomed. Eng. Online* 2020, 19, 9.
61. Hansmann, J., Groeber, F., Kahlig, A., Kleinhans, C., and Walles, H. Bioreactors in tissue engineering—Principles, applications and commercial constraints. *Biotechnol. J.* 2013, 8, 298–307.
62. Ahmed, S., Chauhan, V. M., Ghaemmaghami, A. M., and Aylott, J. W. New generation of bioreactors that advance extracellular matrix modelling and tissue engineering. *Biotechnol. Lett.* 2019, 41, 1–25.
63. Pörtner, R., and Giese, C. *An Overview on Bioreactor Design, Prototyping and Process. Control for Reproducible Three-Dimensional Tissue Culture*, Wiley: Hoboken, NJ, USA, 2006.
64. Hansmann, J., Groeber, F., Kahlig, A., Kleinhans, C., and Walles, H. Bioreactors in tissue engineering—Principles, applications and commercial constraints. *Biotechnol. J.* 2013, 8, 298–307.
65. Salehi-Nik, N., Amoabediny, G., Pouran, B., Tabesh, H., Shokrgozar, M. A., Haghighipour, N., Khatibi, N., Anisi, F., Mottaghy, K., and Zandieh-Doulabi, B. Engineering parameters in bioreactor's design: A critical aspect in tissue engineering. *Biomed. Res. Int.* 2013, 2013, 762132.
66. Goh, J.-Y., Weaver, R. J., Dixon, L., Platt, N. J., and Roberts, R. A. Development and use of in vitro alternatives to animal testing by the pharmaceutical industry 1980–2013. *Toxicol. Res.* 2015, 4, 1297–1307.
67. Kitaeva, K. V., Rutland, C. S., Rizvanov, A. A., and Solovyeva, V. V. Cell culture based in vitro test systems for anticancer drug screening. *Front. Bioeng. Biotechnol.* 2020, 8, 322.
68. Haddadi, M. In vitro ADME screening instead of in vivo studies in preclinical safety. *Biomed. J. Sci. Tech. Res.* 2020, 24, 1837118376.
69. Sharma, L., and Garg, P.K. (Eds.). Artificial Intelligence: Technologies, Applications, and Challenges (1st ed.). Chapman and Hall/CRC, 2021. https://doi.org/10.1201/9781003140351
70. Möller, J., Bhat, K., Riecken, K., Pörtner, R., Zeng, A.-P., and Jandt, U. Process-induced cell cycle oscillations in CHO cultures: Online monitoring and model-based investigation. *Biotechnol. Bioeng.* 2019, 116, 2931–294.
71. Moser, A., Appl, C., Brüning, S., and Hass, V. C. Mechanistic mathematical models as a basis for digital twins. *Adv. Biochem. Eng. Biotechnol.* 2020, 1, 133–180.
72. Abt, V., Barz, T., Cruz-Bournazou, M. N., Herwig, C., Kroll, P., Möller, J., Pörtner, R., and Schenkendorf, R. Model-based tools for optimal experiments in bioprocess engineering. *Curr. Opin. Chem. Eng.* 2018, 22, 244–225.
73. Gargalo, C. L., de las Heras, S. C., Jones, M. N., Udugama, I., Mansouri, S. S., Krühne, U., and Gernaey, K. V. *Towards the Development of Digital Twins for the Bio-Manufacturing Industry*, Springer: Berlin/Heidelberg, Germany, 2021.
74. Kroll, P., Hofer, A., Stelzer, I. V., and Herwig, C. Workflow to set up substantial target-oriented mechanistic process models in bioprocess engineering. *Process. Biochem.* 2017, 62, 24–36.
75. Morris, P. D., Narracott, A., von Tengg-Kobligk, H., Soto, D. A. S., Hsiao, S., Lungu, A., Evans, P., Bressloff, N. W., Lawford, P. V., Hose, D. R., et al. Computational fluid dynamics modelling in cardiovascular medicine. *Heart* 2015, 102, 18–28.
76. Berg, P., Saalfeld, S., Voß, S., Beuing, O., and Janiga, G. A review on the reliability of hemodynamic modeling in intracranial aneurysms: Why computational fluid dynamics alone cannot solve the equation. *Neurosurg. Focus* 2019, 47, E15.

77. Jossen, V., Eibl, D., and Eibl, R. Numerical methods for the design and description of in vitro expansion processes of human mesenchymal stem cells. *Adv. Biochem. Eng. Biotechnol.* 2020, 177, 185–228.
78. Hamet, P., and Tremblay, J. Artificial intelligence in medicine. *Metab. Clin. Exp.* 2017, 69, S36–S40.
79. Jokanović, V. Smart Healthcare in Smart Cities. In *Towards Smart World* (1st ed.), Taylor and Francis Group: New York, 2020, pp. 45–73, Chapter 4.
80. Jokanović, V. Brain-Computer Interface: State-of-Art, Challenges and Future. In *Artificial Intelligence Technologies, Applications, and Challenge*, Taylor and Francis, 2021, (25 pages), pp. 141–177.
81. Jokanović, V. Synthetic Biology and Artificial Intelligence, Computer vision. In *Artificial Intelligence Technologies, Applications, and Challenge*, Taylor and Francis, 2021, pp. 193–211.
82. Jokanović, V. State of the Art of Artificial Intelligence in Dentistry and Its Expected Future, Computer vision. In *Artificial Intelligence Technologies, Applications, and Challenge*, Taylor and Francis, 2022, pp. 103–211.
83. Lopez, C., Tucker, S., Salameh, T., and Tucker, C. An unsupervised machine learning method for discovering patient clusters based on genetic signatures. *J. Biomed. Inform.* 2018, 85, 30–39.
84. Petersen, B. K., Yang, J., Grathwohl, W. S., Cockrell, C., Santiago, C., An, G., and Faissol, D. M. Deep reinforcement learning and simulation as a path toward precision medicine. *J. Comput. Biol.* 2019, 26, 597–604.
85. Esteva, A., Robicquet, A., Ramsundar, B., Kuleshov, V., DePristo, M., Chou, K., Cui, C., Corrado, G., Thrun, S., and Dean, J. A guide to deep learning in healthcare. *Nat. Med.* 2019, 25, 24–29.
86. Von Stosch, M., Davy, S., Francois, K., Galvanauskas, V., Hamelink, J.-M., Luebbert, A., Mayer, M., Oliveira, R., O'Kennedy, R., Rice, P., et al. Hybrid modeling for quality by design and PAT-benefits and challenges of applications in biopharmaceutical industry. *Biotechnol. J.* 2014, 9, 719–726.
87. Jokanović, V. The Extraordinary Importance of 6G Network Development and 3D Holography in Future Healthcare. In *Deep Learning in Internet of Things for Next Generation Healthcare*, Taylor and Francis, 2024, p. 18.
88. Jossen, V., Pörtner, R., Kaiser, S.C., Kraume, M., Eibl, D., and Eibl, R. Mass Production of Mesenchymal Stem Cells—Impact of Bioreactor Design and Flow Conditions on Proliferation and Differentiation. In *Cells and Biomaterials in Regenerative Medicine*, Eberli, D. (Ed.), InTech: London, UK, 2014.
89. Jokanović, V. *The Wondrous Challenges of the Exotic Worlds of Nanomachines and Bio-Inspired Artificial Intelligence, Technological Prospects and Social Applications of Society 5.0*, USA: Chapman and Hall/CRC, 2023, p. 41.
90. Zhang, H., Dai, S., Bi, J., and Liu, K.-K. Biomimetic three-dimensional microenvironment for controlling stem cell fate. *Interface Focus* 2011, 1, 792–803.
91. Bratt-Leal, A. M., Carpenedo, R. L., and McDevitt, T. C. Engineering the embryoid body microenvironment to direct embryonic stem cell differentiation. *Biotechnol. Prog.* 2009, 25, 43–51.
92. Chen, A., Ting, S., Seow, J., Reuveny, S., and Oh, S. Considerations in designing systems for large scale production of human cardiomyocytes from pluripotent stem cells. *Stem Cell Res. Ther.* 2014, 5, 12.
93. He, Y.-T., Qi, Y.-N., Zhang, B.-Q., Li, J.-B., and Bao, J. Bioartificial liver support systems for acute liver failure: Asystematicreviewand meta-analysis of the clinical and preclinical literature. *World J. Gastroenterol.* 2019, 25, 3634–3648.

94. Meyer, S., Peters, N., Mann, T., Wolber, R., Pörtner, R., and Nierle, J. In vitro efficacy and release study with anti-inflammatory drugs incorporated in adhesive transdermal drug delivery systems. *J. Pharm. Sci.* 2014, 103, 1142–1148.
95. Jokanović, V. *Nanomedicine, the Greatest Challenge of the 21st Century*, 2012, 3(2). Drug Des ISSN:2169-0138 DDO, an open access journal
96. Geris, L., and Papantoniou, I. The third era of tissue engineering: reversing the innovation drivers. *Tissue Eng. Part A* 2019, 25, 821–826.
97. Campos, D. F., and De Laporte, L. Digitally fabricated and naturally augmented in vitro tissues. *Adv. Healthc. Mater.* 2020, 10, 2001253.
98. Arai, F., Stumpf, P. S., Ikushima, Y. M., et al. Machine learning of hematopoietic stem cell divisions from paired daughter cell expression profiles reveals effects of aging on self-renewal. *Cell Syst.* 2020, 11, 640–652.
99. Cilloni, D., Petiti, J., Campia, V., et al. Transplantation induces profound changes in the transcriptional asset of hematopoietic stem cells: Identification of specific signatures using machine learning techniques. *J. Clin. Med.* 2020, 9, 1670.
100. Yan, X., Sedykh, A., Wang, W., et al. Construction of a web-based nanomaterial database by big data curation and modeling friendly nanostructure annotations. *Nat. Commun.* 2020, 11, 1–10.
101. Rezende, R. A. , Kasyanov, V., Mironov, V., et al. Organ printing as an information technology. *Proc. Eng.*, 2015, 110, 151–158.
102. Dernowsek, J. D., Rezende, R. A., and Lopes da Silva, J. V. The role of information technology in the future of 3D biofabrication. *J. 3D Print Med.* 2017, 1, 63–74.
103. Göttlich, C., Müller, L. C., Kunz, M., Schmitt, F., Walles, H., Walles, T., Dandekar, T., Dandekar, G., and Nietzer, S. L. A Combined 3D tissue engineered in vitro/in silico lung tumor model for predicting drug effectiveness in specific mutational backgrounds. *J. Vis. Exp.* 2016, e53885. https://doi.org/10.3791/53885.
104. Stratmann, A. T., Fecher, D., Wangorsch, G., Göttlich, C., Walles, T., Walles, H., Dandekar, T., Dandekar, G., and Nietzer, S. L. Establishment of a human 3D lung cancer model based on a biological tissue matrix combined with a Boolean in silico model. *Mol. Oncol.* 2014, 8, 351–365.

9 Building Future-Ready Cities

The Power of Digital Twins in Indo-Dutch Collaboration

Sudesh Agrawal, Naman Agrawal, Himanshu Agrawal, and Sonal Agrawal

9.1 INTRODUCTION: A NEW ERA OF URBAN INFRASTRUCTURE

The world is on the cusp of a digital revolution that is transforming the way cities are planned, built, and managed. As urbanization accelerates – by 2050, 68% of the world's population is projected to live in cities (United Nations, 2018) – the need for more efficient, sustainable, and adaptable infrastructure has never been more critical. Enter Digital Twin technology, a groundbreaking innovation poised to reshape urban infrastructure by creating dynamic, real-time digital replicas of physical spaces. By integrating data from various sources – such as IoT devices, sensors, and historical records – Digital Twins enable cities to optimize resources, predict outcomes, and make data-driven decisions that significantly enhance urban living.

Digital Twins provide a platform for simulation, modeling, and prediction that can address a range of urban challenges, from traffic congestion to energy consumption. According to a report by MarketsandMarkets (2020), the global Digital Twin market is expected to grow from USD 3.1 billion in 2020 to USD 48.2 billion by 2026, at a compound annual growth rate (CAGR) of 58%. This rapid adoption reflects the urgency of cities to evolve in the face of unprecedented challenges, such as climate change, population growth, and resource constraints Las Vegas Innovation District. (2022b)..

The concept of Smart Cities – urban areas that use digital technology to enhance performance and well-being, reduce costs, and minimize resource consumption – has gained significant momentum in recent years. Governments worldwide are embracing Smart City solutions to improve urban management. In India, for example, the government has launched the Smart Cities Mission, which aims to promote sustainable and inclusive development by creating 100 smart cities across the country (Ministry of Housing and Urban Affairs, 2021). The rise of Smart Cities has paved the way for the integration of Digital Twins into urban planning and infrastructure

DOI: 10.1201/9781003582489-9

management. By leveraging real-time data and advanced simulations, cities can respond more effectively to the needs of their residents while optimizing resources. A study by ABI Research (2021) predicts that by 2025, over 500 cities globally will adopt Digital Twin platforms to manage urban infrastructure. This movement is further bolstered by advancements in artificial intelligence (AI), machine learning (ML), and the Internet of Things (IoT), which enable Digital Twins to process vast amounts of data with unprecedented speed and accuracy.

Digital Twins are already playing a pivotal role in shaping the cities of tomorrow. Amsterdam, one of the world's leading smart cities, has been at the forefront of using Digital Twin technology to manage urban growth sustainably. The Amsterdam Digital Twin City Project simulates the effects of urban interventions, such as the placement of electric vehicle (EV) charging stations, optimizing public transport routes, and enhancing energy efficiency. By predicting traffic flow and energy demands, Amsterdam has reduced its energy consumption and congestion by up to 10%, contributing to its goal of becoming climate-neutral by 2050 (City of Amsterdam, 2021a).

Similarly, the Port of Rotterdam, Europe's largest port, has deployed a Digital Twin to improve its operations. Using real-time data from IoT sensors, the port's Digital Twin monitors ship movements, weather conditions, and energy consumption, enabling more efficient docking and departure processes. This has resulted in a 20% reduction in ship waiting times and a significant cut in fuel consumption, aligning with the port's ambition to reduce its carbon footprint (Port of Rotterdam Authority, 2022b).

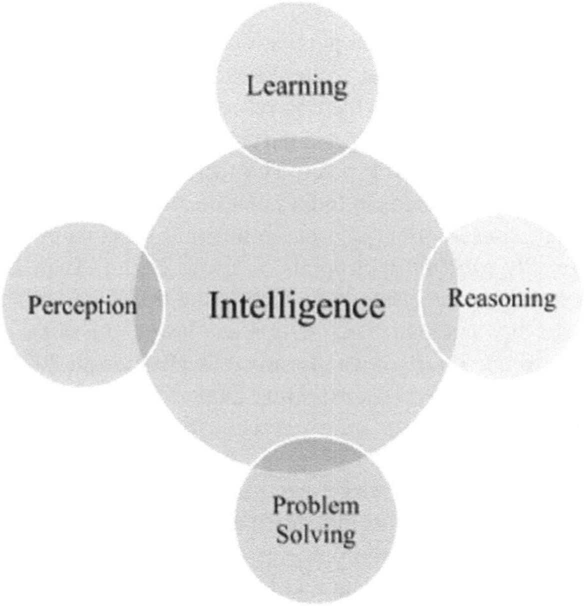

FIGURE 9.1 Subfields of artificial intelligence (AI)

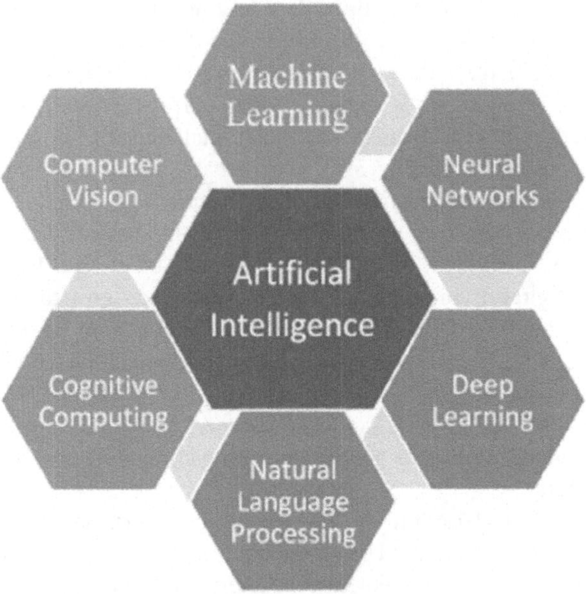

FIGURE 9.2 Various elements of intelligence

9.1.1 The Indo-Dutch Collaboration: A Game Changer for India's Smart Cities

As one of the fastest-growing urban populations in the world, India stands to benefit immensely from Digital Twin technology. The Indo-Dutch collaboration on Digital Twin research and development is set to revolutionize the country's urban infrastructure, particularly through the partnership between IIT Hyderabad and Amsterdam's Digital Twin City Project. This collaboration aims to apply Dutch expertise in real-time simulations to India's unique urban challenges, such as traffic congestion, energy consumption, and environmental sustainability. For instance, Hyderabad, a rapidly growing metropolis, is using Digital Twin models to optimize its traffic management systems. By simulating traffic patterns and analyzing real-time data, the city can reduce congestion and lower emissions, contributing to the broader goals of the Smart Cities Mission (IIT Hyderabad, 2023a, d). As these technologies evolve, the potential for scaling Digital Twin solutions across India's 100 smart cities is immense, promising to reshape the future of urban infrastructure in the country.

Digital Twins are not just the next frontier in urban planning – they are the present. By providing real-time insights into the functioning of cities, Digital Twins offer a powerful tool to make urban environments more efficient, sustainable, and resilient. As global cities like Amsterdam and Hyderabad demonstrate the potential of these technologies, the Indo-Dutch collaboration stands as a beacon of how international partnerships can drive innovation and create the cities of the future.

9.2 DIGITAL TWIN TECHNOLOGY: FROM CONCEPT TO REALITY

Digital Twin technology, once a futuristic concept reserved for the realms of science fiction, has become a transformative force in urban development. At its core, a Digital Twin is a virtual replica of a physical object or system, created by integrating real-time data from various sources, such as sensors, IoT devices, and historical data. This virtual model allows for simulations, analyses, and predictions that offer unprecedented insights into how physical entities behave in real-world conditions. What started as a tool for optimizing manufacturing processes has now evolved into a sophisticated technology reshaping urban infrastructure, energy management, traffic systems, and public services (sharma,l,. 2022).

The concept of Digital Twins was first introduced by Dr. Michael Grieves in 2002 as a theoretical model for product lifecycle management in manufacturing (Grieves & Vickers, 2017). Over the years, the technology matured rapidly, thanks to advancements in computing power, AI, machine learning, and the Internet of Things (IoT). By 2011, NASA was using Digital Twin technology to simulate spacecraft conditions and operations for better decision-making during missions (Shafto et al., 2012). This marked a turning point, showcasing the immense potential of Digital Twins in fields far beyond manufacturing.

As urban populations surge – an additional 2.5 billion people are expected to live in cities by 2050 (United Nations, 2018) – the application of Digital Twins in urban environments has gained traction. Today, cities are turning to Digital Twin models to optimize their operations, improve public services, and develop more sustainable infrastructure. Gartner (2021) estimates that by 2025, 60% of cities globally will leverage Digital Twin platforms to monitor and manage urban environments in real time (sharmaet.al., 2023).

What makes Digital Twin technology truly revolutionary is its ability to process vast amounts of real-time data from a multitude of sources. This data feeds into the digital model, creating a dynamic, living representation of the physical world. For instance, sensors placed across a city's infrastructure – such as buildings, roads, and public utilities – continuously send information to the Digital Twin, which then simulates and predicts future scenarios based on that data.

Take Singapore, for example. The city-state has invested heavily in its Virtual Singapore initiative, a Digital Twin model that provides detailed 3D visualizations of the entire city, from building interiors to transportation networks. With data collected from IoT sensors, Virtual Singapore can simulate scenarios such as population growth, disaster management, and energy consumption. This not only helps urban planners visualize and plan more effectively but also supports real-time decision-making for crisis response (Loh, 2020).

9.2.1 Real-World Applications: From Concept to Citywide Implementation

Several cities around the world have already adopted Digital Twin technology, with each implementing unique solutions tailored to its specific urban challenges. One of the most notable examples is Helsinki's Digital Twin. As part of its carbon-neutrality agenda, Helsinki has developed a Digital Twin of the entire city to simulate how new buildings and infrastructure will impact energy consumption, traffic flow, and carbon emissions. This has enabled the city to optimize its building plans to reduce emissions and improve energy efficiency by up to 25% (City of Helsinki, 2021).

In the United States, the city of Las Vegas is using a Digital Twin to address its growing traffic management problems. By integrating real-time traffic data, the Digital Twin can predict congestion points and adjust traffic signals to optimize vehicle flow. This has resulted in a 15% reduction in traffic delays, significantly improving mobility across the city (Las Vegas Innovation District, 2022a).

Meanwhile, China's Xiong'an New Area, a planned city near Beijing, has been designed from the ground up with Digital Twin technology. The city's infrastructure is embedded with sensors and IoT devices that feed real-time data into a central Digital Twin platform, which manages everything from water distribution to public transportation and energy consumption. Xiong'an's Digital Twin allows the city to operate more efficiently while reducing water waste by 30% and cutting energy usage by 15% (Tao et al., 2019).

The integration of artificial intelligence (AI) and machine learning (ML) is driving Digital Twins from mere virtual replicas to autonomous, self-improving systems. AI and ML enable Digital Twins to analyze large datasets, identify patterns, and make predictive recommendations. This capability allows cities to go beyond monitoring and managing their infrastructure to proactively anticipate problems and optimize systems for future challenges Dutch Ministry of Infrastructure and Water Management. (2022).

In Copenhagen, for example, AI-powered Digital Twins have been used to simulate the effects of climate change on the city's flood management system. By predicting future flood patterns based on real-time data and weather forecasts, the city has been able to design flood defenses that are expected to reduce flood-related damages by 40% (City of Copenhagen, 2022c).

9.2.2 Overcoming Challenges in Digital Twin Adoption

While the potential of Digital Twin technology is immense, there are several challenges to its widespread adoption, particularly in developing countries. One significant barrier is the digital infrastructure required to support the vast amount of data generated by these models. Developing countries often lack the necessary IoT networks, high-speed internet, and cloud computing infrastructure to implement Digital Twins effectively. Furthermore, data privacy concerns and cybersecurity risks must be addressed as cities become more reliant on real-time data.

However, international collaborations, like the Indo-Dutch partnership, are working to bridge these gaps. By combining the Netherlands' advanced expertise in Digital Twin technology with India's ambitious Smart Cities Mission, the Indo-Dutch collaboration is helping Indian cities overcome infrastructural and technological challenges. For instance, the partnership between IIT Hyderabad and Amsterdam's

Digital Twin City Project is laying the groundwork for the deployment of Digital Twins in Indian cities, aiming to improve urban mobility, traffic management, and sustainability (IIT Hyderabad, 2023b).

Digital Twin technology has rapidly progressed from a theoretical concept to a tangible, transformative tool in urban infrastructure. With real-time data at its core and the integration of AI and ML, Digital Twins are revolutionizing how cities are managed, making them smarter, more efficient, and more sustainable. As leading cities like Singapore, Helsinki, and Las Vegas demonstrate the potential of these models, the continued evolution of Digital Twins will redefine urban living, helping cities meet the challenges of the 21st century head-on.

9.3 CASE STUDY 1: AMSTERDAM'S DIGITAL TWIN CITY PROJECT

Amsterdam has emerged as a global leader in harnessing Digital Twin technology to transform urban spaces, pioneering the use of fully integrated, data-driven approaches to urban planning and management. The Amsterdam Digital Twin City Project exemplifies the city's commitment to using real-time data, advanced simulations, and predictive modeling to address key challenges such as traffic congestion, energy efficiency, and sustainability. Through these innovations, Amsterdam aims to achieve its ambitious goal of becoming a climate-neutral city by 2050.

Amsterdam's vision of becoming climate-neutral by 2050 has been a driving force behind the development of its Digital Twin. Policymakers recognized that real-time data and predictive models would be crucial to achieving this goal. Designed not just as a monitoring tool, the city's Digital Twin simulates future scenarios and optimizes urban infrastructure. The City of Amsterdam's smart city development office notes that the platform is central to reducing energy consumption, improving air quality, and enhancing public transportation systems (City of Amsterdam, 2021a, b).

At its core, the Amsterdam Digital Twin integrates vast amounts of data from IoT sensors, traffic cameras, and environmental monitoring systems. This data feeds into a virtual replica of the city, allowing planners to simulate urban interventions such as new building construction, changes in transportation routes, or the installation of renewable energy infrastructure.

9.3.1 Enhancing Traffic Management with Real-Time Data

One of the most impactful uses of Amsterdam's Digital Twin has been in traffic management. Like many cities, Amsterdam faces significant traffic congestion, particularly during peak hours. The Digital Twin has helped alleviate these issues by simulating traffic patterns and making real-time adjustments to optimize vehicle flow.

By analyzing data from traffic cameras and sensors, the Digital Twin can predict congestion points and suggest alternative routes. In one notable case, Amsterdam used its Digital Twin to simulate rerouting traffic from busy intersections, resulting in a 12% improvement in traffic flow during peak hours. Following this success, the strategy was rolled out citywide in 2022 (van den Burg et al., 2022).

The system is also integrated with Amsterdam's public transportation network to optimize bus and tram routes. By simulating passenger flow and traffic conditions, the Digital Twin adjusts schedules to minimize delays, reducing public transport delays by 15% and enhancing mobility and sustainability (City of Amsterdam, 2022).

Beyond traffic management, Amsterdam's Digital Twin is instrumental in advancing the city's energy efficiency and sustainability initiatives. As part of its goal to reduce carbon emissions by 55% by 2030, Amsterdam uses the Digital Twin to simulate energy-saving measures. For instance, it helps city planners identify optimal locations for electric vehicle (EV) charging stations by analyzing traffic density, proximity to residential areas, and access to renewable energy.

Through these simulations, over 500 new EV charging stations were installed in 2021, with plans to reach 1,000 by 2025 (Amsterdam Smart City, 2021). This initiative has contributed to a 20% increase in EV adoption, aligning with Amsterdam's sustainability goals.

The Digital Twin is also used to improve energy efficiency in buildings. By integrating real-time data on energy consumption, weather patterns, and building occupancy, the platform simulates retrofitting measures like improved insulation or energy-efficient lighting. In 2021, the city implemented such measures in public buildings, leading to an 18% reduction in municipal energy consumption (van der Meer et al., 2021).

9.3.2 Predicting the Future: Scenario Planning and Urban Development

A key advantage of Amsterdam's Digital Twin is its ability to simulate the long-term effects of urban planning decisions. For example, when city planners considered building a new residential district in the western part of the city, the Digital Twin was used to predict impacts on traffic, energy consumption, and public services. The simulation revealed that, without infrastructure upgrades, the development would increase traffic congestion by 15%. As a result, the city expanded bike lanes, added bus routes, and built energy-efficient housing, ensuring that the project aligned with sustainability goals (Amsterdam Smart City, 2022a).

The Digital Twin has also been invaluable for climate change preparedness. Given Amsterdam's vulnerability to flooding, the city has used its Digital Twin to simulate flood scenarios and evaluate the effectiveness of flood prevention strategies. By testing dike reinforcements and urban drainage systems, the city has developed a flood defense strategy expected to reduce flood risk by 30% over the next decade (Rijkswaterstaat, 2022).

Looking ahead, Amsterdam's Digital Twin will continue to evolve, with plans to integrate AI and machine learning to enable more autonomous decision-making. This could allow the system to adjust traffic signals, optimize energy distribution, and suggest urban development strategies in real time. Additionally, the city plans to enhance the platform's 3D modeling capabilities, offering an immersive virtual environment for city planners to visualize interventions in greater detail.

As Amsterdam's Digital Twin City Project grows, it serves as a model for other cities worldwide, showcasing the power of data and simulation to address modern urban challenges. From traffic management to energy efficiency and climate resilience, Amsterdam's experience highlights the immense potential of Digital Twin technology in creating smarter, more sustainable cities for future generations.

9.4 CASE STUDY 2: IIT HYDERABAD AND HYDERABAD'S SMART CITY VISION

Hyderabad, one of India's fastest-growing cities, is a leader in the Smart Cities Mission, an initiative launched by the Government of India in 2015 to develop 100 smart cities nationwide. By embracing Digital Twin technology, Hyderabad is tackling challenges posed by rapid urbanization, traffic congestion, and resource management. Central to this effort is IIT Hyderabad, a leading research institution that plays a key role in developing smart city solutions, particularly through its collaboration with the Greater Hyderabad Municipal Corporation (GHMC).

IIT Hyderabad has become a crucial partner in the city's smart city journey, leveraging its research in Digital Twin technology to optimize urban mobility, energy consumption, and public services. Given Hyderabad's projected population growth – expected to reach 15 million by 2030 – the adoption of these technologies is critical to managing the city's expanding infrastructure (United Nations, 2021).

Hyderabad's partnership with Amsterdam's Digital Twin City Project has provided insights into how advanced simulations can address urban challenges, underscoring the city's ambition to lead in smart urban planning and technology integration.

9.4.1 Urban Mobility: Tackling Traffic Congestion with Digital Twins

Traffic congestion is a major issue in Hyderabad, where the number of registered vehicles doubled from 3 million in 2015 to 6 million in 2023 (GHMC, 2023). To combat this, Hyderabad implemented a Digital Twin-based traffic management system developed in partnership with IIT Hyderabad. This system uses real-time data from IoT sensors to monitor traffic conditions and predict congestion.

In 2022, a pilot project in the Madhapur IT corridor – a high-traffic area – used the Digital Twin to simulate traffic patterns and adjust signal timings. This resulted in a 25% reduction in congestion during peak hours (IIT Hyderabad, 2022a). Following this success, the system was expanded to other parts of the city, improving vehicle flow and reducing emissions by 10% (GHMC, 2023).

Energy consumption in Hyderabad has surged due to population growth, prompting the city to focus on energy efficiency in public infrastructure. The GHMC and IIT Hyderabad deployed a Digital Twin model to monitor energy usage in real time, leading to significant savings.

In 2021, Hyderabad installed smart streetlights that adjust brightness based on traffic data from the Digital Twin, reducing energy consumption for street lighting by 30%, saving the city INR 150 million annually (GHMC, 2022). The Digital Twin also helped optimize energy use in public buildings, reducing energy consumption by 20% in retrofitted government buildings (IIT Hyderabad, 2023d).

9.4.2 Scaling Digital Twin Technology across India

Hyderabad's Digital Twin-based water management system, developed with IIT Hyderabad, integrates real-time data from sensors in reservoirs, pipelines, and drainage systems to optimize water distribution and prepare for water shortages or floods. During a 2022 drought, the system ensured efficient water distribution, while in the monsoon season, it helped implement flood prevention measures that reduced flood-related damages by 15% (IIT Hyderabad, 2022a, b).

In addition to infrastructure management, Hyderabad's Digital Twin technology is enhancing public services like waste management. Smart waste bins with sensors monitor fill levels and send data to the Digital Twin, optimizing collection routes and reducing fuel consumption. In 2023, this system cut waste management costs by 20% and improved collection efficiency by 30% (GHMC, 2023). The Digital Twin is also used for emergency response, simulating disaster scenarios to help emergency services allocate resources effectively.

Hyderabad's success in implementing Digital Twin technology has made it a model for other Indian cities. As part of the Indo-Dutch partnership, Hyderabad collaborates with cities like Pune, Chennai, and Bengaluru to share lessons and expand Digital Twin systems across India's Smart Cities Mission. IIT Hyderabad is providing technical expertise to accelerate this process, helping cities across the country adopt smart, sustainable urban planning.

Hyderabad's adoption of Digital Twin technology, driven by its collaboration with IIT Hyderabad, has transformed its approach to urban planning and infrastructure management. From reducing traffic congestion to improving energy efficiency and water management, Digital Twin technology has proven to be a powerful tool for enhancing the city's sustainability and resilience. As Hyderabad continues to lead in India's Smart Cities Mission, it serves as an inspiring example of how technology and innovation can create smarter, more livable cities for the future.

9.5 SUSTAINABILITY THROUGH DIGITAL TWINS: LESSONS FROM THE NETHERLANDS

As cities worldwide grapple with the increasing pressures of climate change, resource scarcity, and rapid urbanization, the Netherlands has emerged as a global leader in the application of Digital Twin technology to promote sustainability. Known for its innovation in environmental management and smart urban planning, the Netherlands is leveraging Digital Twins to optimize energy consumption, reduce carbon emissions, and enhance the efficiency of public services. These advancements offer valuable lessons for cities around the world, particularly for those in developing nations like India, where urbanization and environmental challenges are accelerating.

In this section, we explore how the Netherlands has applied Digital Twin technology in groundbreaking ways, focusing on key projects such as the Port of Rotterdam, the Living Lab Scheveningen, and national-scale initiatives in energy management and public services. These examples highlight how cities can harness real-time data, simulations, and predictive modeling to build more sustainable and resilient urban environments.

9.5.1 The Port of Rotterdam: A Model for Sustainable Logistics

The Port of Rotterdam, Europe's largest seaport, is a prime example of how Digital Twin technology can drive sustainability in urban and industrial environments. As a critical hub for global trade, the Port of Rotterdam handles over 470 million tons of cargo annually, making it a vital artery for European and international commerce (Port of Rotterdam Authority, 2022a, b). However, with this scale of operation comes significant environmental challenges, including high levels of energy consumption, emissions, and logistical inefficiencies.

To address these issues, the Port of Rotterdam Digital Twin was developed in collaboration with the Rotterdam Maritime Capital of Europe and various technology partners. The Digital Twin integrates real-time data from sensors placed across the port, including weather stations, shipping data, and energy usage information. This data is fed into a virtual replica of the port, enabling operators to monitor conditions in real time, simulate various operational scenarios, and optimize processes to reduce environmental impact.

One of the key successes of the Port of Rotterdam's Digital Twin is its ability to reduce ship waiting times and energy consumption. By simulating ship movements and predicting optimal docking schedules, the Digital Twin has helped reduce the average waiting time for vessels by 20% (Port of Rotterdam Authority, 2022a, b). This has led to a corresponding decrease in fuel consumption, as ships spend less time idling in the port, contributing to a 15% reduction in carbon emissions from maritime traffic.

In addition to improving logistics, the Digital Twin has been instrumental in optimizing the port's energy usage. By analyzing energy consumption data from cranes, warehouses, and other port infrastructure, the Digital Twin has identified inefficiencies and recommended changes, such as retrofitting equipment with energy-efficient technologies. These measures have resulted in a 10% reduction in overall energy consumption at the port, supporting the Netherlands' broader goal of achieving carbon neutrality by 2050 (Port of Rotterdam Authority, 2022b).

9.5.2 The Living Lab Scheveningen: A Smart Coastal City

Another pioneering project in the Netherlands is the Living Lab Scheveningen, located along the coastline of The Hague. This smart city initiative uses Digital Twin technology to improve the sustainability and efficiency of public services in the coastal district of Scheveningen. The Living Lab integrates real-time data from IoT sensors, environmental monitoring systems, and public utilities to create a virtual model of the area, allowing city planners and service providers to optimize operations.

One of the key applications of the Digital Twin in Scheveningen is in energy management. The coastal district has installed a network of smart streetlights that are connected to the Digital Twin. These streetlights automatically adjust their brightness based on real-time data from traffic sensors and environmental conditions, such as daylight levels and weather forecasts. This has resulted in a 35% reduction in energy consumption for street lighting, contributing to the city's overall sustainability goals (City of The Hague, 2022).

In addition to energy management, the Living Lab Scheveningen is using its Digital Twin to enhance waste management and public safety. Smart waste bins equipped with sensors monitor fill levels and send real-time data to the Digital Twin, which then optimizes waste collection routes to reduce fuel consumption and operational costs. The system has improved waste collection efficiency by 25% and reduced emissions from waste collection vehicles by 15% (City of The Hague, 2022).

The Digital Twin is also being used to improve public safety in Scheveningen by integrating data from surveillance cameras, traffic sensors, and emergency response systems. For example, during large public events, such as festivals or beach gatherings, the Digital Twin can simulate crowd movements and traffic flow, allowing city authorities to manage crowds more effectively and ensure public safety. This has led to a 20% reduction in response times for emergency services in the district (City of The Hague, 2022).

9.5.3 Scaling Sustainability: National Initiatives in Energy and Public Services

Beyond individual city projects, the Netherlands has made significant strides in scaling Digital Twin technology to promote sustainability on a national level. The Dutch government, in collaboration with private sector partners and academic institutions, has developed several national-scale initiatives that leverage Digital Twins to optimize energy usage, reduce emissions, and improve public services across the country (city of copenhagen 2022b, 2022c).

One such initiative is the National Energy Twin, a Digital Twin model developed to support the country's transition to renewable energy. The model integrates real-time data from wind turbines, solar panels, energy grids, and weather stations to simulate the production and distribution of renewable energy across the Netherlands. By analyzing this data, the National Energy Twin can predict energy demand and supply imbalances, enabling grid operators to optimize energy distribution and prevent outages.

In 2021, the National Energy Twin played a crucial role in balancing energy supply during a particularly windy season, when wind turbines generated an excess of electricity. The Digital Twin simulated various energy storage and distribution scenarios, helping grid operators divert excess energy to storage facilities and reduce reliance on fossil fuel-based power plants. This initiative contributed to a 12% reduction in the country's carbon emissions from energy production during the year (Dutch Ministry of Economic Affairs and Climate Policy, 2022).

Another national initiative is the Digital Twin for Public Health Services, which was developed in response to the COVID-19 pandemic. This Digital Twin integrates real-time data from hospitals, testing centers, and vaccination sites to monitor the spread of the virus and optimize healthcare resource allocation. By simulating various pandemic scenarios, the model helped the Dutch government predict hospital capacity needs, manage vaccine distribution, and implement targeted lockdown measures. As a result, the country was able to reduce the strain on its healthcare system and minimize the economic impact of the pandemic (National Institute for Public Health and the Environment, 2021).

9.5.4 Lessons for Global Sustainability Efforts

The Netherlands' success in using Digital Twin technology to promote sustainability offers several valuable lessons for cities and countries around the world. First and foremost, the Dutch experience demonstrates the importance of integrating real-time data into urban management. By collecting and analyzing data from IoT sensors, environmental monitoring systems, and public utilities, cities can create dynamic Digital Twins that allow for more informed decision-making and more effective resource management.

Second, the Netherlands has shown that collaboration between public and private sectors is key to scaling Digital Twin technology. Many of the country's most successful Digital Twin projects, such as the Port of Rotterdam and the Living Lab Scheveningen, have been developed through partnerships between government agencies, technology companies, and research institutions. This collaborative approach has allowed the Netherlands to harness cutting-edge technology and apply it to real-world challenges in a way that benefits both the public and private sectors.

Third, the Dutch experience highlights the importance of flexibility and adaptability in the use of Digital Twins. Whether managing energy grids, optimizing traffic flow, or improving public safety, Digital Twin models must be able to adapt to changing conditions and respond to new data in real time. This requires continuous investment in technology infrastructure, such as IoT networks and cloud computing, as well as a commitment to data privacy and cybersecurity.

9.5.5 A Blueprint for the Future

The Netherlands' innovative use of Digital Twin technology provides a blueprint for cities and countries looking to build more sustainable, efficient, and resilient urban environments. By integrating real-time data, simulations, and predictive modeling, Dutch cities like Rotterdam and The Hague have demonstrated the transformative potential of Digital Twins in areas such as logistics, energy management, and public services.

As the world faces increasing environmental challenges, the lessons learned from the Netherlands offer hope and inspiration for how technology can be harnessed to create a more sustainable future. For cities like Hyderabad, which are already beginning to implement Digital Twin solutions as part of their smart city initiatives, the Netherlands serves as a model for how technology, innovation, and collaboration can drive meaningful change on a global scale.

9.6 CONCLUSION: DIGITAL TWINS AS CORNERSTONES OF SMART CITY INFRASTRUCTURE

As cities around the world face unprecedented challenges, from rapid urbanization to the impacts of climate change, Digital Twin technology is emerging as a critical tool in shaping the future of urban infrastructure. By integrating real-time data, artificial intelligence (AI), and predictive analytics, Digital Twins offer city planners, policymakers, and engineers a dynamic, powerful way to manage urban environments more effectively. Throughout this chapter, we have explored how cities like Amsterdam, Hyderabad, and Copenhagen are leveraging Digital Twins to revolutionize traffic management, enhance energy efficiency, and promote sustainability. The success of these initiatives provides a roadmap for how Digital Twins can become the cornerstones of smart city infrastructure globally.

One of the most significant advantages of Digital Twins is their ability to provide data-driven insights into the complex systems that make up a city. Traditional urban planning relied heavily on static models and historical data, which often failed to account for the dynamic and interconnected nature of modern cities. Digital Twins, however, offer a real-time, evolving view of urban systems, from transportation networks to energy grids and water management systems. This shift toward data-driven urban planning enables cities to be more responsive, efficient, and resilient in the face of growing challenges.

Take, for example, Amsterdam's Digital Twin City Project, which integrates data from thousands of IoT sensors across the city. This real-time data allows the city to simulate and predict the effects of various urban interventions, such as changes to traffic flow, the installation of electric vehicle (EV) charging stations, or the development of new housing. By providing a dynamic and constantly updated model of the city, Amsterdam's Digital Twin has helped reduce traffic congestion by 12% and energy consumption in public buildings by 18%, contributing significantly to the city's goal of becoming climate-neutral by 2050 (City of Amsterdam, 2021a, b).

9.6.1 Enhancing Sustainability and Resilience

As cities strive to meet global climate goals, sustainability and resilience have become central to urban development. Digital Twins are playing a crucial role in helping cities reduce their environmental footprint by optimizing the use of energy, water, and other resources. For instance, Rotterdam's Digital Twin has significantly reduced the port's carbon emissions by 15% by optimizing ship docking schedules and minimizing energy use across the port's infrastructure (Port of Rotterdam Authority, 2022a). These kinds of improvements are not only helping cities meet their climate targets but also ensuring that they remain resilient in the face of environmental changes.

Moreover, Digital Twins enable cities to simulate future scenarios, allowing them to prepare for natural disasters and other emergencies. The ability to model the impacts of climate change, such as rising sea levels or increased flooding, allows cities to build more resilient infrastructure. In Copenhagen, for example, the city's Digital Twin has been used to simulate the effects of climate change on its flood management system. By modeling future flood patterns, Copenhagen has been able to design flood defenses that are expected to reduce flood-related damages by 40% (City of Copenhagen, 2022c). This kind of predictive modeling is crucial for cities that are increasingly vulnerable to climate change.

The potential of Digital Twin technology goes beyond individual cities; it has the power to scale and transform urban environments globally. The Indo-Dutch collaboration has demonstrated how international partnerships can play a key role in scaling these solutions. By sharing expertise, resources, and technological innovations, India and the Netherlands have been able to apply Digital Twin technology in cities like Hyderabad and Chennai, addressing challenges related to traffic congestion, water management, and energy efficiency (IIT Hyderabad, 2023c, d).

These international collaborations serve as a model for how countries around the world can work together to implement Digital Twin technologies. By adopting the lessons learned from cities like Amsterdam and Rotterdam, urban centers in developing countries can leapfrog traditional urban planning approaches and deploy cutting-edge solutions that promote sustainability and resilience. Moreover, organizations like the World Bank and the United Nations are increasingly recognizing the role of Digital Twins in helping cities meet the Sustainable Development Goals (SDGs), particularly in areas like clean energy, climate action, and sustainable cities (World Bank, 2021 & 2022).

9.6.2 The Path Forward: Future of Digital Twins in Urban Development

Looking ahead, the role of Digital Twins in urban development is only set to grow. With advancements in AI, machine learning, and cloud computing, Digital Twins will become even more powerful, enabling cities to simulate and manage increasingly complex systems. Autonomous vehicles, for example, will rely heavily on Digital Twin models to navigate urban environments safely and efficiently. As more cities adopt smart technologies, Digital Twins will be essential for ensuring that urban infrastructure can keep pace with technological advancements.

Furthermore, the integration of citizen engagement into Digital Twin platforms offers exciting possibilities. By allowing residents to interact with digital models of their cities, governments can make urban planning processes more transparent and inclusive. Cities like Singapore are already experimenting with such initiatives, allowing citizens to visualize and provide feedback on urban projects through the city's Digital Twin (Singapore Land Transport Authority, 2022).

Digital Twins are becoming indispensable tools for the future of urban infrastructure. By providing real-time insights, optimizing resource use, and enhancing resilience, they are helping cities navigate the complex challenges of the 21st century. The success stories from Amsterdam, Rotterdam, and Hyderabad highlight the transformative potential of this technology, while international collaborations like the Indo-Dutch partnership underscore the importance of scaling these solutions globally.

As cities continue to grow and evolve, the integration of Digital Twins into urban planning will be key to building smarter, more sustainable, and more resilient cities. In a world where urban populations are expected to rise dramatically in the coming decades, the ability to manage and optimize urban systems in real time will be crucial. By positioning Digital Twins as the cornerstones of smart city infrastructure, cities around the world can create a brighter, more sustainable future for generations to come.

REFERENCES

ABI Research. (2021). *Smart cities and digital twin technology report.* Available at: https://www.abiresearch.com [Accessed on 22 October 2024].

Amsterdam Smart City. (2021). *Amsterdam's EV charging station strategy: Data-driven deployment for sustainability.* Available at: https://www.amsterdamsmartcity.com [Accessed on 22 October 2024].

City of Amsterdam. (2021a). *Digital twin city: Amsterdam's vision for the future.* Available at: https://www.amsterdam.nl [Accessed on 22 October 2024].

City of Amsterdam. (2021b). *Digital twin technology and climate-neutrality by 2050.* Available at: https://www.amsterdam.nl [Accessed on 22 October 2024].

City of Amsterdam. (2022). *Optimizing public transport with digital twins: A case study.* Available at: https://www.amsterdam.nl [Accessed on 22 October 2024].

City of Copenhagen. (2022a). *Copenhagen City Digital Twin: Enhancing flood resilience and sustainability.* Available at: https://www.copenhagen.dk [Accessed on 22 October 2024].

City of Copenhagen. (2022b). *Copenhagen City Digital Twin: Optimizing traffic and reducing emissions.* Available at: https://www.copenhagen.dk [Accessed on 22 October 2024].

City of Copenhagen. (2022c). *Flood management and digital twin technology.* Available at: https://www.copenhagen.dk [Accessed on 22 October 2024].

City of Helsinki. (2021). *Carbon-neutrality and digital twin technology for urban planning.* Available at: https://www.helsinki.fi [Accessed on 22 October 2024].

City of The Hague. (2022). *Living Lab Scheveningen: Smart coastal city initiatives.* Available at: https://www.denhaag.nl [Accessed on 22 October 2024].

Dutch Ministry of Economic Affairs and Climate Policy. (2022). *National Energy Twin: Balancing renewable energy supply and demand.* Available at: https://www.government.nl [Accessed on 22 October 2024].

Dutch Ministry of Infrastructure and Water Management. (2022). *Water management and digital twin solutions: Indo-Dutch collaboration in Chennai.* Available at: https://www.government.nl [Accessed on 22 October 2024].

Gartner. (2021). *Market forecast for digital twin technology in smart cities*. Available at: https://www.gartner.com [Accessed on 22 October 2024].

GHMC. (2022). *Smart street lighting and energy efficiency: Hyderabad's digital twin approach*. Greater Hyderabad Municipal Corporation.

GHMC. (2023). *Traffic management and waste collection optimization through digital twins: Hyderabad case study*. Greater Hyderabad Municipal Corporation.

Grieves, M., & Vickers, J. (2017). Digital twin: Mitigating unpredictable, undesirable emergent behaviour in complex systems. In F.-J. Kahlen, S. Flumerfelt, & A. Alves (Eds.), *Transdisciplinary perspectives on complex systems* (pp. 85–113). Springer.

IIT Hyderabad. (2022a). *Digital twin pilot project for traffic management in Hyderabad's IT corridor*. Available at: https://www.iith.ac.in [Accessed on 22 October 2024].

IIT Hyderabad. (2022b). *Scaling digital twin energy solutions: Smart streetlights in Hyderabad*. Available at: https://www.iith.ac.in [Accessed on 22 October 2024].

IIT Hyderabad. (2023a). *Energy efficiency in public buildings using digital twins: Hyderabad's smart city vision*. Available at: https://www.iith.ac.in [Accessed on 22 October 2024].

IIT Hyderabad. (2023b). *Indo-Dutch partnership in smart cities and digital twin technology*. Available at: https://www.iith.ac.in [Accessed on 22 October 2024].

IIT Hyderabad. (2023c). *Indo-Dutch partnership in smart city development and digital twin technologies*. Available at: https://www.iith.ac.in [Accessed on 22 October 2024].

IIT Hyderabad. (2023d). *Smart Cities Mission and digital twin technology research*. Available at: https://www.iith.ac.in [Accessed on 22 October 2024].

Las Vegas Innovation District. (2022a). *Autonomous vehicles and AI-driven traffic management: Case study from Las Vegas*. Available at: https://www.lasvegasnevada.gov [Accessed on 22 October 2024].

Las Vegas Innovation District. (2022b). *Digital twin technology in traffic management: A success story*. Available at: https://www.lasvegasnevada.gov [Accessed on 22 October 2024].

Loh, D. (2020). *Virtual Singapore: Building a digital twin for the city-state*. Government Technology Agency. Available at: https://www.gov.sg [Accessed on 22 October 2024].

MarketsandMarkets. (2020). *Digital twin market forecast to 2026*. Available at: https://www.marketsandmarkets.com [Accessed on 22 October 2024].

Ministry of Housing and Urban Affairs. (2021b). *Smart Cities Mission: A national initiative for urban transformation*. Government of India.

National Institute for Public Health and the Environment. (2021). *Digital twin for public health services: Optimizing pandemic response*. Available at: https://www.rivm.nl [Accessed on 22 October 2024].

Port of Rotterdam Authority. (2022a). *Digital twin enhances port efficiency*. Available at: https://www.portofrotterdam.com [Accessed on 22 October 2024].

Port of Rotterdam Authority. (2022b). *Sustainability and digital twin technology: Reducing emissions in Europe's largest port*. Available at: https://www.portofrotterdam.com [Accessed on 22 October 2024].

Rijkswaterstaat. (2022). *Flood prevention strategies: The role of digital twin technology*. Available at: https://www.rijkswaterstaat.nl [Accessed on 22 October 2024].

Shafto, M., et al. (2012). *Modelling, simulation, information technology, and processing roadmap*. NASA.

Sharma, L., & Carpenter, M. (Eds.). (2022). Computer Vision and Internet of Things: Technologies and Applications (1st ed.). Chapman and Hall/CRC. https://doi.org/10.1201/9781003244165.

Sharma, L., & Garg, P.K. (Eds.). (2023). *Technological Prospects and Social Applications of Society 5.0* (1st ed.). Chapman and Hall/CRC. https://doi.org/10.1201/9781003324720

Singapore Land Transport Authority. (2022). *Digital twin for citizen engagement in urban planning*. Available at: https://www.lta.gov.sg [Accessed on 22 October 2024].

Tao, F., et al. (2019). Digital twin-driven product design framework. *International Journal of Production Research*, 57(12), 3935–3953.

United Nations. (2018). *World urbanization prospects: The 2018 revision*. New York: United Nations. [Accessed on 22 October 2024].

United Nations. (2021). *World urbanization prospects: The 2021 revision*. New York: United Nations.

Van den Burg, R., van der Meer, P., & de Jong, H. (2022). Traffic management and digital twin technology: Case studies from Amsterdam. *Journal of Smart City Planning*, 14(2), 45–62.

Van der Meer, P., et al. (2021). Energy efficiency in public buildings using digital twin simulations: Amsterdam's approach. *Sustainable Cities Journal*, 10(3), 22–35.

World Bank. (2021). *Smart city solutions and sustainable urban development: Lessons from India and the Netherlands*. Available at: https://www.worldbank.org [Accessed on 22 October 2024].

World Bank. (2022). *Scaling water management solutions in South Asia through international partnerships*. Available at: https://www.worldbank.org [Accessed on 22 October 2024].

10 A Transductive and Inductive GNNs for Physical Moving Objects Detection in Surface Scenes for Digital Twins

Wieke Prummel, Jhony Giraldo, Badri N. Subudhi, Anastasia Zakharova, and Thierry Bouwmans

10.1 INTRODUCTION

To reflect physical moving objects accurately in digital twins, their detection in videos is an essential process in order to track and recognize them in various environments like in surface surveillance [1], bio-diversity surveillance [2–4], maritime surveillance [5, 6], ocean surveillance [7–9], and underwater surveillance [10–13]. Surface and underwater environments present their own specific challenges such as illumination variations and dynamic backgrounds [14–17]. To handle the surface scene challenges, different machine learning models have been employed and can be classified into two main categories: (a) representation learning [18–20] and incremental robust subspace learning [21–23], and (b) neural networks learning [24–26]. In the last decade, deep neural networks [27–30] have shown strong performance on large-scale datasets, as the quality of data improves, these models become increasingly complex and computationally intensive, even with fast algorithms [31], few-shot learning methods [32], and specialized architectures [33–36]. In this chapter, we focus on transductive GNNs [37–41] and inductive GNNs [42] approaches as they require less labeled data and they are the most practicable for digital twins. The rest of this chapter is organized as follows: A short overview of MOD models based on deep neural networks is provided in Section 10.2. Transductive and inductive GNN approaches are reviewed in Section 10.3 and Section 10.4, respectively. Section 10.5 presents a comparative evaluation on a large-scale CDnet 2014 dataset which contains videos of surface environments. Finally, conclusions and future works are given in Section 10.6.

DOI: 10.1201/9781003582489-10

10.2 PHYSICAL MOVING OBJECTS DETECTION: A SHORT PRELIMINARY OVERVIEW

Moving Objects Detection (MOD) can be viewed as a labeling task with pixels classified as background (BG) or foreground (FG) by comparing the background image with the current image. MOD is an important technique in computer vision before recognition and tracking physical moving objects [43] and thus MOD is suitable for digital twins. MOD can be divided into three main groups: mathematical models, machine learning models, and signal processing models. Statistic [44], fuzzy [45], and Dempster-Shäfer [46] methods allow to manage imprecision, uncertainty, and incompleteness in the data, related to different challenges and to the quality of the sensors. About robust representation learning models, unsupervised techniques look for low-rank structure in the data to be robust against outliers in matrix [47–51] or tensor representations [52–56]. Deep Learning models attempt to face all the challenges by learning all the situations [57–59] and are more efficient than methods based on traditional hand-crafted features [60] and feature selection schemes [61]. Signal processing models estimate the background based on the temporal distribution of the pixels in image frames [62–64]. In practice, deep learning-based methods are dominating the performance on the CDnet 2014 datasets with six models: (1) FgSegNet_M [65] and its variants FgSegNet_S [66] and FgSegNet_V2 [67], (2) BSGAN [68] and its variant BSPVGAN [69], and (3) Cascaded CNNs [70]) for supervised approaches. However, background subtraction is a classification task and can be solved successfully with DNNs. But their performance decreases dramatically in unseen scenes being scenes specific. Moreover, they require a large amount of labeled data not always available in several applications. In this context, graph signal processing approaches [38–41] offer a good compromise between performance and practicability in unseen scenarios. Based on this fact, several GNN-based methods have been developed in literature and they can be classified as follows:

- Transductive approaches do not adhere to a strict separation of training and testing phases; rather, they integrate both labeled and unlabeled data simultaneously. Transductive learning algorithms include graph-based methods such as Label Propagation [71–73], self-training techniques, and semi-supervised approaches like Generative Adversarial Networks (GANs) [74]. As shown in Figure 10.1, one single graph is built from the full dataset. The graph vertices are then subdivided into training, testing, and validation sets.
- Inductive approaches distinctly separate the training and testing sets, as opposed to transductive learning methods. This approach has been widely applied in tasks such as image classification, object detection, and semantic segmentation. Notable algorithms for inductive learning in computer vision include Convolutional Neural Networks (CNNs) [75], and decision trees [76]. By learning from labeled examples, inductive learning enables models to generalize effectively to new instances, making it a foundational approach in computer vision research. Figure 10.1 shows also an inductive scenario in which multiple graphs are built instead of a single one. This

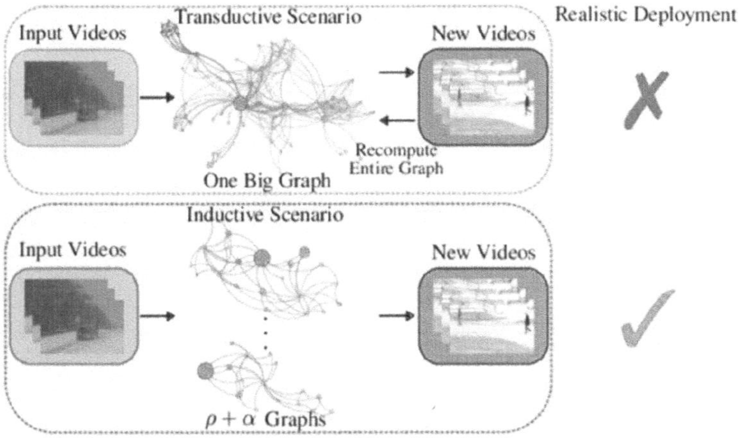

FIGURE 10.1 Transductive versus inductive learning for physical MOD

strategy avoids the need to rebuild the full graph and retrain the model when a new video is incorporated into the dataset, thereby enhancing its suitability for real-world deployments.

10.3 TRANSDUCTIVE GRAPH MOD APPROACH

GraphMOS method was proposed in 2020 by Giraldo et al. [38]. This method introduces an approach to MOD using graph-based techniques by addressing various challenges in the field. Graphs are a powerful data representation that models complex data and expresses intrinsic relationships between different data entities. As demonstrated by Shuman et al. [77] and Sandryhaila et al. [78], the concept of graph signal processing (GSP) extends classical signal processing to graphs, enabling graph filtering operations on nodes. One of the principal motivations behind GraphMOS is the utilization of graph signal processing (GSP) for the reconstruction of graph signals from labeled samples. GraphMOS is the first work applying graph signal sampling and reconstruction techniques to the field of MOS and video object segmentation (VOS). However, for the sake of brevity, we will limit our discussion to the MOS application. The theoretical advances are as follows: The paper makes two significant theoretical contributions: (1) it provides an upper bound on the sample complexity necessary for semi-supervised learning under certain prior conditions, and (2) two bounds for the condition number of a perturbed matrix are established. GraphMOS is subjected to a comprehensive evaluation process, with its performance being benchmarked against state-of-the-art methods on four widely used MOS datasets. The study presents a comparative analysis of the algorithms' applicability to both static and moving camera sequences.

GraphMOS improves the accuracy and efficiency of moving object segmentation in video sequences by classifying each pixel region as a static or moving object through a semi-supervised graph signal reconstruction approach. The algorithm

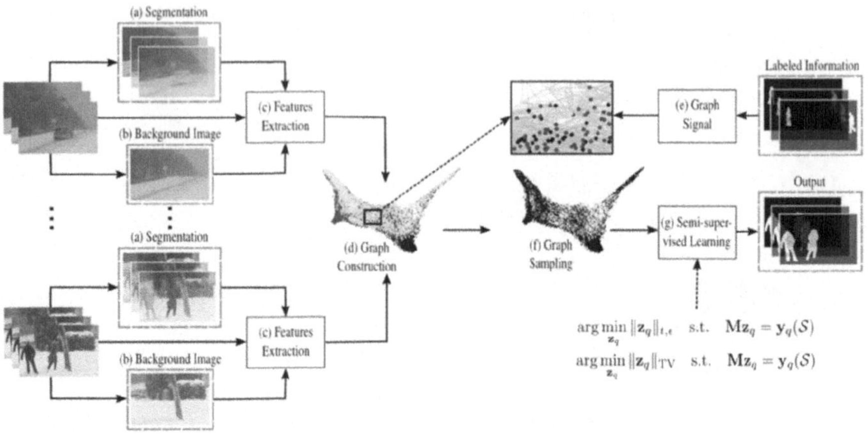

FIGURE 10.2 The pipeline of GraphMOS

comprises several steps: segmentation, background initialization, feature extraction, graph signal attribution, graph signal sampling, graph construction, and a semi-supervised learning recovery method. Figure 10.2 illustrates the framework of the GraphMOS algorithm, which includes the reconstruction of graph signals. The algorithm begins with background initialization and superpixel segmentation. In this context, each superpixel is treated as a node in a graph, and the characteristics of each node are derived from motion, intensity, texture, and deep features. Ground truth (GT) data are employed to determine whether a node represents a moving object (indicated by green nodes) or a static object (indicated by blue nodes). The black nodes denote images within the dataset that remain unlabeled. Finally, a subset of nodes is chosen for labeling, and the semi-supervised algorithm proceeds to reconstruct the labels for all nodes in the graph.

10.4 INDUCTIVE GRAPH MOD APPROACH

GraphIMOS (Graph-based Inductive Moving Object Segmentation) [42] is an inductive graph-based framework, which provides a significant advancement in the practical application of graph-based methods in video surveillance. Indeed, a Graph Convolutional Network (GCN) architecture was developed with a specific focus on its suitability for MOS. GraphIMOS is the first inductive learning methodology employing graphs for MOS. The workflow is provided in Figure 10.3. In this approach, each video instance is represented as a node within a graph framework. To construct these graphs, a Mask Region Convolutional Neural Network (Mask R-CNN) [79] featuring a ResNet-*50* backbone and a Feature Pyramid Network (FPN) was employed. The nodes were characterized using intensity, optical flow, and texture features. As an alternative to constructing a single, extensive graph, $\rho + \alpha$ *k*-Nearest Neighbors (*k*-NN) graphs were generated, each comprising a distinct number of nodes. For the training and validation graphs, ρ was set to 3 and they were fed into the proposed

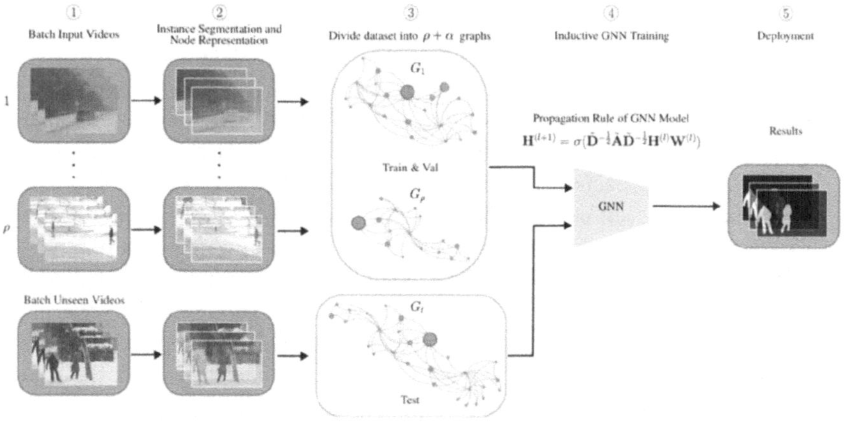

FIGURE 10.3 Pipeline of GraphIMOS

Graph Neural Network (GNN) model for training and hyperparameter tuning. In the inductive learning scenario, it is essential that the nodes and edges in the test data do not overlap with those in the training data. In order to evaluate the effectiveness of the approach, an in-depth analysis was conducted on a single graph instance $\alpha = 1$, constructed from previously unseen videos. GraphIMOS underwent comprehensive testing on the challenging CDnet 2014 dataset [80], where competitive performance was shown compared to existing inductive methods.

10.4.1 GraphIMOS: An Overview

The GraphIMOS architecture is provided in Figure 10.3 and consists of the following steps: instance segmentation, graph construction, and the implementation of a Graph Convolutional Network (GCN).

- **Instance Segmentation and Node Labeling:** The Mask R-CNN [79] with the ResNet-*50* backbone is employed to identify regions, utilizing a *k*-NN technique to construct graphs. The feature representation framework proposed by Giraldo et al. [38] is chosen, wherein the outputs from Mask R-CNN serve as nodes in the constructed graphs. The aforementioned nodes are represented by means of a concatenation of all extracted features, which results in a high-dimensional vector (*930*-dimensional) that serves as a unique descriptor for each instance. To assign labels to the detected instances, each node in the graph is associated with a graph signal represented by $y \in \mathbb{R}$. Two different labels can be assigned to the detected instances: moving or static objects. Here, N corresponds to the number of detected instances. Each row of the graph signal matrix $Y \in \mathbb{R}^{N \times Q}$ contains a one-hot encoded vector that corresponds to the class label of the object in the segmented region. Specifically, $Q = 2$ is the number of classes, thus *1* is used to denote moving objects, while *0* represents static objects.

- **Graph Mini-batch Construction:** The proposed framework represents a significant departure from previous graph-based moving object segmentation (MOS) methods, as it employs an inductive architectural approach. In contrast to approaches like GraphMOS [38], where the entire graph must be rebuilt for each newly added video in the dataset, necessitating the re-solving of the optimization problem and complicating real-world deployment, our approach avoids this by generating $\rho + \alpha$ separate graphs, each with a different number of nodes. This methodology guarantees that the data within the ρ training-validation graphs remain unconnected. To build a comprehensive graph that includes multiple distinct subgraphs, the adjacency matrices are organized in a block-diagonal fashion as follows:

$$A = \begin{bmatrix} A_1 & 0- & 0 \\ 0 & \cdots & | \\ | & & 0 \\ 0 & -0 & A_\rho \end{bmatrix}$$

In this context, the following definitions are applicable: A represents the adjacency matrix in the mini-batch; ρ denotes the number of training and validation graphs; and α represents the testing graphs, as depicted in Figure 10.3. The adjacency matrices are stored efficiently using a sparse representation that only records non-zero entries, thus avoiding extra memory overhead. Furthermore, node and target features are seamlessly integrated into the node dimension through simple concatenation.

- **Implementation of the GCN Model:** The model has been constructed with two GCNConv layers, as described by [81], multiple ReLU activation layers, and three linear layers. To mitigate the issue of overfitting and enhance the model's ability to generalize to unseen videos, five dropout layers and four PairNorm batch normalization layers were incorporated into the model [82]. These techniques also contribute to the stability of the training process and prevent over-smoothing. PairNorm is especially advantageous for facilitating more effective model convergence. The primary objective is to learn meaningful representations of the graph structure and node features for the purpose of classifying objects as either static or moving. The final layer of our model employs a log softmax function, which is defined as follows:

$$\log \text{softmax} = \log(\frac{1}{s} \exp(x_i)), \text{ where } s = \sum_s \exp(x_i)).$$

The proposed GCN-based architecture is implemented using the PyTorch Geometric library [83]. The k-NN graphs are constructed with k=*40* neighbors. To prevent overfitting, dropout is applied with a coefficient of *0.5*. Stochastic gradient descent (SGD) is employed for optimization, with a momentum of *0.9*, a learning rate of *0.01*, and a weight decay of 5×10^{-4}. The model is trained for a maximum of *500* epochs. The architecture comprises five hidden layers, and the negative log likelihood loss function is employed for training and evaluation of the GCN-based model.

10.4.2 Experimental Setup for GraphIMOS

The efficacy of GraphIMOS is assessed in comparison to a number of contemporary algorithms utilizing the extensive CDnet 2014 dataset [80]. To establish an inductive framework, multiple graphs were constructed by dividing the data into four sequences. As illustrated in Table 10.1, the four sequences are designated as S1, S2, S3, and S4. Consequently, four experiments are conducted to compute the F-measure for the entire dataset. These sequences have been selected in order to facilitate evaluation of unseen videos according to the approach outlined in [84]. For each sequence, the results are computed and a graph is built for that sequence. The four graphs are designated as G1, G2, G3, and G4. In each experimental trial, two graphs are employed for the purposes of training and evaluation with regard to GraphIMOS, while a third graph is utilized for the testing phase. In order to test the algorithm's performance on unseen videos, each experiment uses a distinct graph for testing purposes.

Sequences	Training	Validation	Testing
S2,S3,S1,S4 (Exp1)	*G2, G3*	*G1*	*G4*
S1,S3,S4,S2 (Exp2)	*G1, G3*	*G4*	*G2*
S2,S3,S4,S1 (Exp3)	*G2, G3*	*G4*	*G1*
S1,S2,S4,S3 (Exp4)	*G1, G2*	*G4*	*G3*

10.5 EXPERIMENTAL RESULTS

To have a fair comparison, we present the results obtained on the publicly available CDnet 2014 [80] dataset. The dataset comprises 22 additional camera-captured videos providing 5 different categories, compared to CDnet 2012. Those additional videos are used to incorporate some challenging issues that were not addressed in the CDnet 2012 dataset. The categories are listed as follows: baseline, dynamic or complex backgrounds, camera jitter, shadowing, intermittent object motion, thermal, bad weather, low frame rate, night videos, PTZ, and turbulence or storminess. In the CDnet 2014, the ground truths (GT) of only the first half of every video in the five new categories are made publicly available for testing, other than CDnet 2012 which publishes the GT of all video frames. However, the evaluation is reported for all frames. The F-measures obtained by transductive and inductive GNNs algorithms are compared with the F-measures of other representative deep learning algorithms over the complete evaluation dataset. In addition, we indicate if algorithms fall into the following categories: pixel-wise algorithms, spatial-wise algorithms, and temporal-wise. On the other hand, UCSD contains 18 videos mainly composed of moving

TABLE 10.1

F-Measure Metric for the Categories of CDnet 2014 Dataset, Namely Baseline (BSL), Dynamic background (DBG), Camera jitter (CJT), Intermittent Motion Object (IOM), Shadows (SHD), Thermal (THM), Bad Weather (BDW), Low Frame Rate (LFR), Night Videos (NVD), PTZ, Turbulence (TBL)

Algorithms (Authors)	BSL	DBG	CJT	IOM	SHD	THM	BDW	LFR	NVD	PTZ	TBL	Average F-Measure
Transductive Multi-scale/cascaded CNNs												
FgSegNet -M (*Spatial-wise*) [65]	0.9973	**0.9958**	0.9954	0.9951	0.9937	0.9921	0.9845	0.8786	0.9655	0.9843	0.9648	0.9770
FgSegNet-S (*Spatial-wise*) [66]	0.9977	**0.9958**	**0.9957**	0.9940	0.9927	0.9937	0.9897	0.8972	0.9713	**0.9879**	0.9681	0.9804
FgSegNet-V2 (*Spatial-wise*) [67]	**0.9978**	0.9951	0.9938	**0.9961**	**0.9955**	**0.9938**	**0.9904**	**0.9336**	**0.9739**	0.9862	**0.9727**	**0.9847**
Transductive 3D CNNs												
3D CNN (*Temporal-wise*) [85]	0.9691	0.9614	0.9396	**0.9698**	0.9706	0.9830	0.9509	0.8862	0.8565	0.8987	0.8823	0.9507
3D Atrous CNN (*Spatial/Temporal-wise*) [86]	0.9897	**0.9789**	0.9645	0.9637	0.9813	0.9833	0.9609	0.8994	0.9489	0.8582	0.9488	0.9615
FC3D (*Spatial/Temporal-wise*) [87]	0.9941	0.9775	0.9651	0.8779	0.9881	0.9902	0.9699	0.8575	0.9595	0.9240	0.9729	0.9524
MFC3D (*Spatial/Temporal-wise*) [87]	**0.9950**	0.9780	**0.9744**	0.8835	**0.9893**	**0.9924**	**0.9703**	**0.9233**	**0.9696**	**0.9287**	**0.9773**	**0.9619**
Transductive GANs												
BGAN (*Pixel-wise*) [68]	0.9814	0.9763	0.9828	0.9366	0.9849	0.9064	0.9465	0.8472	0.8965	0.9194	0.9118	0.9339
BPVGAN (*Pixel-wise*) [69]	0.9837	**0.9849**	**0.9893**	0.9366	**0.9927**	**0.9764**	0.9644	0.8508	0.9001	**0.9486**	0.9310	0.9501
Transductive Graph Neural Networks												
GraphMOS [38]	0.9398	0.7334	0.7005	0.3607	**0.9653**	**0.7292**	0.8294	**0.5538**		0.7599		0.7302

(Continued)

TABLE 10.1 (CONTINUED)

F-Measure Metric for the Categories of CDnet 2014 Dataset, Namely Baseline (BSL), Dynamic background (DBG), Camera jitter (CJT), Intermittent Motion Object (IOM), Shadows (SHD), Thermal (THM), Bad Weather (BDW), Low Frame Rate (LFR), Night Videos (NVD), PTZ, Turbulence (TBL)

Algorithms (Authors)	BSL	DBG	CJT	IOM	SHD	THM	BDW	LFR	NVD	PTZ	TBL	Average F-Measure
GraphMOD-Net (Original) [37]	**0.9550**	**0.8510**	**0.7200**	**0.5540**	0.9420	0.6820	**0.8390**	0.5210		**0.7700**		**0.7593**
Inductive CNNs/Graph Neural Networks												
FgSegNet (Modified) [65]	0.5641	0.2067	0.2845	0.3325	0.3809	0.3584	0.2789	0.2115		0.1400		0.3061
GraphMOD-Net (Modified) [37]	0.6474	**0.6254**	0.4926	0.5243	0.6587	0.5484	0.6268	0.5337		0.5899		0.5831
GraphIMOS [42]	**0.7003**	0.5868	**0.6700**	**0.5284**	**0.6807**	**0.6453**	**0.6377**	**0.5478**		**0,5932**		**0.6211**

camera sequences, with 30 up to 246 frames for each sequence. Table 10.1 shows the different F-measures which come either from the corresponding papers, or the CDnet 2014 website. In the same way, Figures 10.4 and Figure 10.5 show some visual results obtained using GraphMOD-Net [37], GraphMOS [38], and GraphIMOS [42].

Qualitative Results: GraphMOS [38] provides optimal visual results, as shown in Figure 10.4. In comparison to GraphMOD-Net [37], it has been shown to outperform the sequence featuring adverse weather conditions with wet snow. We can note that for the bird sequences from UCSD which features dynamic background variations such as rippling water surfaces, both GraphMOS and GraphMOD-Net are able to effectively address these challenges. Figure 10.5 shows the visual results

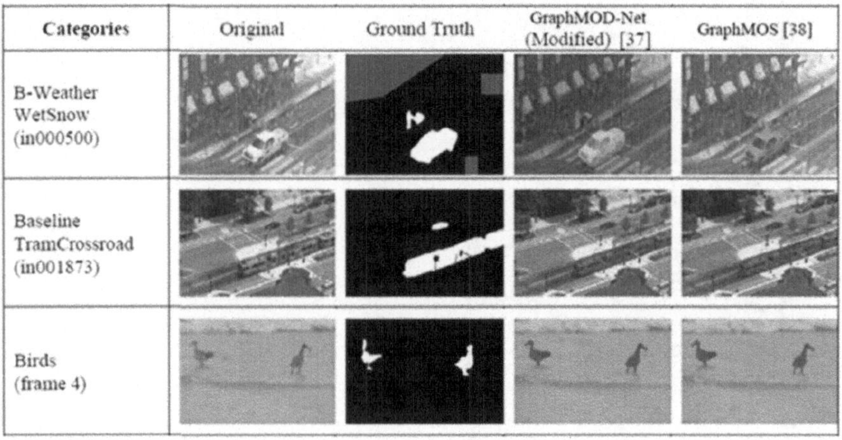

FIGURE 10.4 Visual results of transductive-based methods on CDnet 2014 and UCSD datasets. From left to right: Original images, Ground-Truth images, GraphMOD-Net [45], and GraphMOS [46]

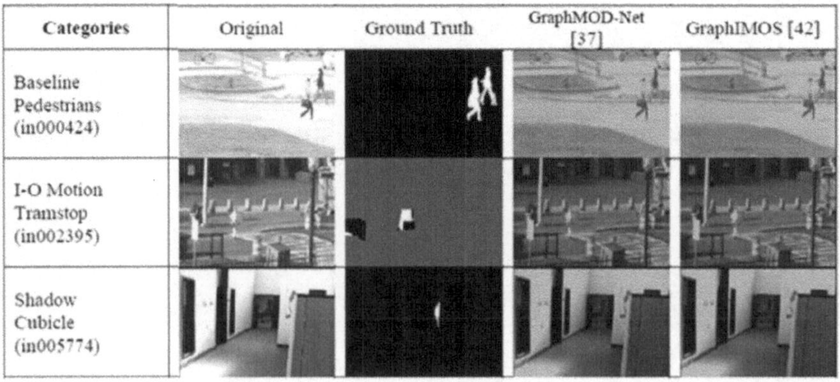

FIGURE 10.5 Visual results of inductive-based methods on CDnet 2014 dataset. From left to right: Original images, Ground-Truth images, GraphMOD-Net [45], and GraphIMOS [50]

of GraphIMOS on the CDnet 2014 dataset. The results show that GraphIMOS outperforms GraphMOD-Net (modified) in the context of intermittent object motion, as evidenced by its superior performance in the tram stop sequence, and in the context of shadow detection, as evidenced by its superior performance in the cubicle scene. The results of the segmentation are markedly superior for GraphIMOS.

Quantitative Results: The assessment of a number of algorithms for MOS has provided some intriguing insights across a range of categories. The results in Table 10.1 suggest that transductive multi-scale/cascaded CNNs may offer a promising approach. Among the models tested, FgSegNet-V2 (spatial-wise) showed a promising performance, achieving an average F-measure of 98%. Additionally, both FgSegNet-M (spatial-wise) and FgSegNet-S (spatial-wise) yielded promising results, with average F-measures of 97.7% and 98%, respectively. These results suggest that multi-scale and cascaded approaches may be effective for MOS tasks. The results suggest that transductive 3D CNNs may have potential for temporal modeling, with the 3D CNN (temporal-wise) [83] achieving an average F-measure of 95%. The 3D Atrous CNN (Spatial/Temporal-wise) [84] showed a slight improvement with an average F-Measure of 96.15%, while the FC3D (Spatial/Temporal-wise) [85] and MFC3D (Spatial/Temporal-wise) [85] also demonstrated promising results, with average F-Measures of 95.24% and 96.19%, respectively. It seems that the MFC3D (Spatial/Temporal-wise) method may offer the most promising results among the transductive 3D CNNs. It appears to perform particularly well on baseline (99.5%), camera jitter (97.44%), and shadow (98.93%). It is worth noting that the model performs particularly well on the thermal, bad weather, low frame rate, night videos, PTZ, and turbulence datasets, with average F-Measures of 99.24%, 97.03%, 96.96%, 97.73%, 92.87%, and 92.33%, respectively. In the dynamic background challenge, the MFC3D (Spatial/Temporal-wise) model achieved a slightly lower result than the 3D Atrous CNN (Spatial/Temporal-wise) model, with a difference of only 0.09%. In the IntermittentObjectMotion challenge, the model's performance was not as strong as that of the other transductive 3D CNN models. Transductive GANs also demonstrated promising results, with BScGAN (Pixel-wise) [86] and BGAN (Pixel-wise) [87] achieving average F-Measures of 97,63% and 93,39%, respectively, and BPVGAN (Pixel-wise) [88] performing well with an average F-Measure of 95,01%. In comparison, transductive graph neural networks did not perform as well as CNN-based approaches. It might be suggested that GraphMOS and GraphMOD-Net (Original) could benefit from further improvement, with average F-Measures of 0.7302 and 0.7593, respectively, indicating room for enhancement in graph-based transductive learning for MOS. Nevertheless, GraphMOD-Net in contrast to GraphMOS uses a GCN model for solving the GraphMOS graph signal reconstruction problem in a semi-supervised manner. Although GraphMOS already demonstrated interesting results, GraphMOD-Net outperforms GraphMOS on baseline (95.5%), dynamic background (85.1%), camera jitter (72%), intermittent object motion (55.4%), bad weather (83.9%), and PTZ (77%). This demonstrates that the GCN model enhances the results of GraphMOS. As anticipated, inductive convolutional neural networks (CNNs) and graph neural networks (GNNs) demonstrated inferior performance compared to their transductive counterparts. This is attributed to the inherent challenges associated with inductive learning, particularly in terms of generalization to unseen data. The modified FgSegNet exhibited markedly inferior performance, with an average F-Measure of 0.3061. The modified GraphMOD-Net model demonstrated an improvement in performance with an average F-Measure of 0.5831. It is noteworthy that GraphIMOS achieved the highest average F-Measure among inductive methods, at 0.6211. This shows its effectiveness in handling unseen video sequences and indicates its potential for real-world applications.

10.6 CONCLUSION

This chapter presents a comprehensive review of recent advancements in the use of GNNs applied to physical moving objects detection in video captured by a static or non-moving camera in surface environments. Experimental outcomes on the large-scale CDnet 2014 dataset show the interest of GNNs compared to the supervised deep neural network methods. However, researchers need to answer the question: What is the best suitable type of graph neural networks and its corresponding architecture for background initialization, BGS, and deep learned features in the presence of dynamic backgrounds? It seems that the current graph neural networks tested meet problems in the case of moving or non-static cameras which will be more powerful for digital twins. Thus, inductive graph methods provide a suitable framework for generating virtual representation of physical moving objects accurately in digital twins.

REFERENCES

1. B. Garcia-Garcia, T. Bouwmans, and A. Rosales-Silva, "Background Subtraction in Real Applications: Challenges, Current Models and Future Directions," *Computer Science Review*, 35, 100204, February 2020.
2. W. Huang, Q. Zeng, and M. Chen, "Motion characteristics estimation of animals in video surveillance," *IEEE Advanced Information Technology, Electronic and Automation Control Conference, IAEAC 2017*, p. 1098–1102, 2017.
3. B. Weinstein, "Scene-specific convolutional neural networks for video-based biodiversity detection," *Methods in Ecology and Evolution*, 2018
4. E. Sheehan, D. Bridger, S. Nancollas, and S. Pittman, "PelagiCam: a novel underwater imaging system with computer vision for semi-automated monitoring of mobile marine fauna at offshore structures," *Environmental Monitoring and Assessment*, 2020
5. Y. Chan, "Comprehensive comparative evaluation of background subtraction algorithms in open sea environments," *Computer Vision and Image Understanding*, 2020
6. Y. Chan, "Ensemble learning-based method for maritime background subtraction in open sea environments," *Computer Vision and Image Understanding*, 238, 103859, January 2024.
7. D. Prasad, C. Prasath, D. Rajan, L. Rachmawati, E. Rajabally, and C. Quek, "Challenges in Video based Object Detection in Maritime Scenario using Computer Vision," *WASET International Journal of Computer, Electrical, Automation, Control and Information Engineering*, 11, No. 1, January 2017

8. D. Prasad, D. Rajan, L. Rachmawati, E. Rajabally, and C. Quek, "Video Processing from Electro-optical Sensors for Object Detection and Tracking in Maritime Environment: A Survey," *IEEE Transactions on Intelligent Transportation Systems*, 1993–2016
9. D. Prasad, C. Prasath, D. Rajan, L. Rachmawati, E. Rajabally, and C. Quek, "Object Detection in a Maritime Environment:Performance Evaluation of Background Subtraction Methods," *IEEE Transactions on Intelligent Transportations Systems*, 1787–1802, No. 5, May 2019
10. F. El Baf, and T. Bouwmans. "Comparison of background subtraction methods for a multimedia learning space," in *International Conference on Signal Processing and Multimedia*, SIGMAP, July 2007.
11. V. Bajpai, et al., "Underwater U-Net: Deep Learning with U-Net for Visual Underwater Moving Object Detection," in *OCEANS 2021*, San Diego, USA, pp. 1–4, September 2021.
12. M. Kapoor, et al., "Underwater Moving Object Detection using an End-to-End Encoder-Decoder Architecture and GraphSage with Aggregator and Refactoring," in *Proceedings of the IEEE/CVF Conference on Computer Vision and Pattern Recognition*, CVPRW, pp. 5636–5645, 2023.
13. M. Nissar, B. Subudhi, V. Jakhetiya, and A. Mishra, "Underwater Change Detection using Multiple Sampling-based Probabilistic Learner and Feature Preservance Discriminator," in *IEEE International Conference on Image Processing, ICIP 2024*, Abu Dhabi, UAE, October 2024.
14. W. Kim, and C. Jun, "Illumination-Invariant Background Subtraction: Comparative Review, Models, and Prospects", IEEE Access, June 2017.
15. S. Sanches, C. Oliveira, A. Sementille, and V. Freire, "Challenging situations for background subtraction algorithms," *Applied Intelligence*, 49, pp. 1771–1784, 2019. https://doi.org/10.1007/s10489-018-1346-4
16. T. Bouwmans, "Traditional and Recent Approaches in Background Modeling for Foreground Detection: An Overview," *Computer Science Review*, Vol. 11–12, pp. 31–66, May 2014.
17. M. Chapel, and T. Bouwmans, "Moving Objects Detection with a Moving Camera: A Comprehensive Review," *Computer Science Review*, Vol. 38, p. 100310, November 2020.
18. S. Javed, A. Mahmood, T. Bouwmans, and S. Jung, "Superpixels based Manifold Structured Sparse RPCA for Moving Object Detection," in *BMVC 2017*, September 2017.
19. S. Javed, A. Mahmood, S. Al-Maadeed, T. Bouwmans, and S. Jung, "Moving object detection in complex scene using spatiotemporal structured-sparse RPCA," in *IEEE T-IP*, Vol. 28, No. 2, pp. 1007–1022, 2019.
20. M. Sultana, A. Mahmood, T. Bouwmans, and S. Jung, "Complete Moving Object Detection in the Context of Robust Subspace Learning," in *ICCV 2019*, Seoul, South Korea, October 2019.
21. P. Narayanamurthy, and N. Vaswai, "A Fast and Memory-efficient Algorithm for Robust PCA (MEROP)," in *IEEE ICASSP 2018*, April 2018.
22. N. Vaswani, T. Bouwmans, S. Javed, and P. Narayanamurthy, "Robust PCA and Robust Subspace Tracking: A Comparative Evaluation," in *IEEE SSP 2018*, Freiburg, Germany, June 2018.
23. T. Bouwmans, N. Vaswani, P. Rodriguez, R. Vidal, and Z. Lin, "Introduction to the Special Issue on Robust Subspace Learning and Tracking: Theory, Algorithms, and Applications," *IEEE Journal of Selected Topics in Signal Processing*, Vol. 12, No.

6, pp. 1127–1130, December 2018. https://doi.org/10.1109/JSTSP.2018.2879245. keywords: {Special issues and sections;Principal component analysis;Data analysis;Signal processing algorithms;Computer vision;Clustering algorithms}
24. L. Maddalena, and A. Petrosino, "Multivalued Background/Foreground Separation for Moving Object Detection," in *WILF 2009*, Vol. 5571, pp. 263–270, Palermo, Italy, June 2009.
25. L. Maddalena, and A. Petrosino, "The 3DSOBS+ algorithm for moving object detection," in *Computer Vision and Image Understanding, CVIU 2014*, May 2014.
26. M. Chacon-Murguia, and S. Gonzalez-Duarte, "An Adaptive Neural-Fuzzy Approach for Object Detection in Dynamic Backgrounds for Surveillance Systems," in *IEEE T-IE*, 2011.
27. M. K. Panda, B. N. Subudhi, T. Bouwmans, V. Jakheytiya and T. Veerakumar, "An End to End Encoder-Decoder Network with Multi-scale Feature Pulling for Detecting Local Changes From Video Scene," 2022 18th IEEE International Conference on Advanced Video and Signal Based Surveillance (AVSS), Madrid, Spain, 2022, pp. 1–8. https://doi.org/10.1109/AVSS56176.2022.9959141. keywords: {Deep learning;Surveillance;Transfer learning;Neural networks;Lighting;Object detection;Feature extraction}.
28. M. Panda, B. Subudhi, T. Bouwmans, V. Jakhetiya, and T. Veerakumar, "An End to End Encoder-Decoder Network with Multi-scale Feature Pulling for Detecting Local Changes From Video Scene," in *IEEE AVSS 2022*, Madrid, Spain, November 2022.
29. Lavanya Sharma, D. K. Yadav, "Histogram based Adaptive Learning Rate for Background Modelling and Moving Object Detection in Video Surveillance", *International Journal of Telemedicine and Clinical Practices,* Inderscience, June 2016, Vol. 2 No.1, pp. 74–92.
30. M. Sultana, A. Mahmood, T. Bouwmans, and S. Jung, "Dynamic Background Subtraction Using Least Square Adversarial Learning," in *IEEE ICIP 2020*, Abu Dhabi, UAE, October 2020.
31. B. Hou, Y. Liu, and N. Ling, "A super-fast deep network for moving object detection," in *IEEE ISCAS 2020*, pp. 1–5, 2020.
32. I. Osman, and M. Shehata, "Few-shot learning network for moving object detection using exemplar-based attention map," in *IEEE ICIP 2022*, pp. 1056–1060, 2022.
33. T. Minematsu, A. Shimada, and R. Taniguchi, "Rethinking background and foreground in deep neural network-based background subtraction," in *IEEE ICIP*, pp. 3229–3233, 2020.
34. J. Giraldo, H. Le, and T. Bouwmans, "Deep Learning based Background Subtraction: A Systematic Survey," in *Handbook of "Pattern Recognition and Computer Vision"*, WSP, March 2020.
35. T. Bouwmans, S. Javed, M. Sultana, and S. Jung, "Deep Neural Network Concepts in Background Subtraction: A Systematic Review and A Comparative Evaluation," *Neural Networks*, Vol. 117, pp. 8–66, 2019.
36. M. Mandal, and S. Vipparthi, "An Empirical Review of Deep Learning Frameworks for Change Detection: Model Design, Experimental Frameworks, Challenges and Research Needs," *IEEE Transactions on Intelligent Transportation Systems*, Vol. 23, No. 7, pp. 6101–6122, July 2022. https://doi.org/10.1109/TITS.2021.3077883. keywords: {Deep learning;Training;Task analysis;Data models;Computational modeling;Cameras;Benchmark testing;Change detection;survey;background subtraction;deep learning;scene independence}2021.
37. J. Giraldo, S. Javed, N. Werghi, and T. Bouwmans, "Graph CNN for Moving Object Detection in Complex Environments from Unseen Videos," in *ICCVW 2021*, October 2021.

38. J. Giraldo, S. Javed, and T. Bouwmans, "Graph Moving Object Segmentation," *IEEE Transactions on Pattern Analysis and Machine Intelligence*, Vol. 44, No. 5, pp. 2485–2503, 1 May 2022. https://doi.org/10.1109/TPAMI.2020.3042093. keywords: {Videos;Task analysis;Signal processing algorithms;Object segmentation;Semisupervised learning;Deep learning;Complexity theory;Moving object segmentation;graph signal processing;semi-supervised learning;unseen videos;video object segmentation}.
39. J. Giraldo, and T. Bouwmans, "Semi-Supervised Background Subtraction of Unseen Videos: Minimization of the Total Variation of Graph Signals," in *ICIP 2020*, Abu Dhabi, UAE, 2020.
40. J. Giraldo, and T. Bouwmans, "GraphBGS: Background Subtraction via Recovery of Graph Signals," in *International Conference on Pattern Recognition, ICPR 2020*, January 2021.
41. J. Giraldo, S. Javed, M. Sultana, S. Jung, and T. Bouwmans, "The Emerging Field of Graph Signal Processing for Moving Object Segmentation," in *IW-FCV 2021*, South Korea, February 2021.
42. W. Prummel, J. Giraldo, A. Zakharova, and T. Bouwmans, "Inductive Graph Neural Networks for Moving Object Segmentation," in *IEEE, ICIP 2023*, Kuala Lumpur, Malaysia, October 2023.
43. Lavanya Sharma, N. Lohan, "Performance Analysis of Moving Object Detection using BGS Techniques", *International Journal of Spatio-Temporal Data Science, Inderscience*, Vol. 1 No.1, pp. 22–53, Feb, 2019 (ISSN: 2399-1275).
44. T. Bouwmans, F. El Baf, and B. Vachon, "Statistical Background Modeling for Foreground Detection: A Survey," in *Handbook of "Pattern Recognition and Computer Vision"*, World Scientific Publishing, Vol. 4, Part 2, Chapter 3, pp. 181–199, January 2010.
45. T. Bouwmans, "Background Subtraction for Visual Surveillance: A Fuzzy Approach," *Handbook on "Soft Computing for Video Surveillance"*, Edited by S. Pal, A. Petrosino and L. Maddalena, CRC Press, Taylor and Francis Group, Chapter 5, March 2012.
46. O. Munteanu, T. Bouwmans, E. Zahzah, and R. Vasiu, "The detection of moving objects in video by background subtraction using Dempster-Shafer theory," *Transactions on Electronics and Communications*, 60, No. 1, 1–9, 2015.
47. T. Bouwmans, and E. Zahzah, "Robust PCA via Principal Component Pursuit: A Review for a Comparative Evaluation in Video Surveillance," *Special Issue on Background Models Challenge, Computer Vision and Image Understanding*, 122, 22–34, May 2014.
48. T. Bouwmans, A. Sobral, S. Javed, S. Jung, and E. Zahzah, "Decomposition into Low-rank plus Additive Matrices for Background/Foreground Separation: A Review for a Comparative Evaluation with a Large-Scale Dataset," *Computer Science Review*, 23, 1–71, February 2017.
49. T. Bouwmans, S. Javed, H. Zhang, and Z. Lin, "On the Applications of Robust PCA in Image and Video Processing," *Proceedings of the IEEE*, 1427–1457, 2018. https://doi.org/10.1109/JPROC.2018.2853589. keywords: {Robustness;Computer vision;Principal component analysis;Statistical analysis;Sparse matrices;Image processing;Matrix decomposition;Image processing;medical imaging;robust principal component analysis (RPCA);video processing;3-D computer vision}.
50. L. Huang, and J. Qin, "Fast Dual-Graph Regularized Background Foreground Separation", IEEE SampTA 2023, pages 1-5, New Haven, USA, 2023.
51. Y. Yang, Z. Yang, J. Le, and J. Li, "Nonconvex γ-norm and Laplacian scale mixture with salient map for moving object detection," *Multimedia Tools and Applications*, 83, 26159–26182, 2024. https://doi.org/10.1007/s11042-023-16561-w
52. S. Javed, T. Bouwmans, and S. Jung, "Stochastic Decomposition into Low Rank and Sparse Tensor for Robust Background Subtraction," in *ICDP 2015*, July 2015.

53. A. Sobral, C. Baker, T. Bouwmans, and E. Zahzah, "Incremental and Multi-feature Tensor Subspace Learning applied for Background Modeling and Subtraction," in *International Conference on Image Analysis and Recognition, ICIAR 2014*, October 2014.
54. Q. Luo, M. Yang, W. Li, and M. Xiao, "Multidimensional Data Processing with Bayesian Inference via Structural Block Decomposition," *IEEE Transactions on Cybernetics*, 54, no. 5, pp. 3132–3145, May 2024. https://doi.org/10.1109/TCYB.2023.3234356. keywords: {Tensors;Bayes methods;Matrix decomposition;Hyperspectral imaging;Probabilistic logic;Data processing;Principal component analysis;Bayesian inference;matrix outer product (MOP);tensor completion (TC);tensor decomposition;tensor robust principal component analysis (TRPCA)}
55. H. Cai, Z. Chao, L. Huang, and D. Needell, "Robust Tensor CUR Decompositions: Rapid Low-Tucker-Rank Tensor Recovery with Sparse Corruptions," *SIAM Journal on Imaging Sciences*, 17, No. 1, 225–247, January 2024.
56. M. Amoozegar, M. Akbarizadeh, and T. Bouwmans, "Robust and Efficient FISTA-based Method for Moving Object Detection under Background Movements," *Knowledge-Based Systems*, 294, 111765, June 2024.
57. T. Minematsu, A. Shimada, and R. Taniguchi, "Analytics of deep neural network in change detection," in *IEEE AVSS 2017*, Lecce, Italy, 2017.
58. Sharma, L., & Carpenter, M. (Eds.). (2022). *Computer Vision and Internet of Things: Technologies and Applications* (1st ed.). Chapman and Hall/CRC. https://doi.org/10.1201/9781003244165
59. T. Hamada, T. Minematsu, A. Shimada, F. Okubo, and Y. Taniguchi, "Background Subtraction Network Module Ensemble for Background Scene Adaptation," in *2022 18th IEEE International Conference on Advanced Video and Signal Based Surveillance (AVSS)*, pp. 1–8, 2022.
60. T. Bouwmans, C. Silva, C. Marghes, M. Zitouni, H. Bhaskar, and C. Frelicot, "On the Role and the Importance of Features for Background Modeling and Foreground Detection," *Computer Science Review*, 28, 26–91, May 2018.
61. C. Pacheco, T. Bouwmans, and C. Frelicot, "Superpixel-based online wagging one-class ensemble for feature selection in foreground/background separation," *Pattern Recognition Letters*, 100, 144–151, December 2017.
62. S. Messelodi, C. Modena, N. Segata, and M. Zanin, "A Kalman filter based background updating algorithm robust to sharp illumination changes," in *ICIAP 2005*, Cagliari, Italy, September 2005.
63. K. Naganuma, and S. Ono, "Cross-Frame Foreground Structural Similarity Modeling by Convolutional Sparse Representation," in *IEEE ASC 2023*, pp. 779–783, Taipei, Taiwan, 2023.
64. J. Liu, M. Feng, D. Gu, W. Liu, and X. Xiu, "Moving object detection in gigapixel-level videos using manifold sparse representation," *Multimedia Tools and Applications*, 83, No. 6, 18381–18405, February 2024.
65. L. Lim, and H. Keles, "Foreground Segmentation using a Triplet Convolutional Neural Network for Multiscale Feature Encoding," *Preprint*, January 2018
66. K. Lim, L. Ang, and H. Keles, "Foreground Segmentation Using Convolutional Neural Networks for Multiscale Feature Encoding," *Pattern Recognition Letters*, 2018.
67. K. Lim, L. Ang, and H. Keles, "Learning Multi-scale Features for Foreground Segmentation," *arXiv preprint arXiv:1808.01477*, 2018.
68. W. Zheng, K. Wang, and F. Wang, "Background subtraction algorithm based on Bayesian generative adversarial networks," *Acta Automatica Sinica*, 2018.
69. W. Zheng, K. Wang, and F. Wang, "A novel background subtraction algorithm based on parallel vision and Bayesian GANs," *Neurocomputing*, 394, 178–200, June 2020.

70. Y. Wang, Z. Luo, and P. Jodoin, "Interactive deep learning method for segmenting moving objects," *Pattern Recognition Letters*, 96, 66–75, 2017.
71. J. Suykens, "Deep Restricted Kernel Machines using Conjugate Feature Duality," *Neural Computation*, 29, 2123–2163, 2017.
72. J. Gast, and S. Roth, "Lightweight Probabilistic Deep Networks", Preprint, 2018.
73. Y. Deng, Z. Ren, Y. Kong, F. Bao, Q. Dai, "A Hierarchical Fused Fuzzy Deep Neural Network for Data Classification," in *IEEE T-FS*, Vol. 25, No. 4, pp. 1006–1012, 2017.
74. X. Zhu, Z. Ghahramani, and J. Lafferty. "Semi-supervised learning using gaussian fields and harmonic functions," in *ICML 2003*, pp. 912–919, 2003.
75. I. Goodfellow, J. Pouget-Abadie, M. Mirza, B. Xu, D. Warde-Farley, S. Ozair, and A. Courville, "Generative adversarial nets," in *NeurIPS 2014*, 2014.
76. A. Krizhevsky, I. Sutskever, and G. Hinton. "Imagenet classification with deep convolutional neural networks," *ACM*, 60, No. 6, 84–90, 2017.
77. D. Shuman, S. Narang, P. Frossard, A. Ortega, and P. Vandergheynst, "The emerging field of signal processing on graphs: Extending high-dimensional data analysis to networks and other irregular domains," *IEEE Signal Processing Magazine*, 30, No. 3, 83–98, 2013.
78. A. Sandryhaila, and J. Moura, "Discrete signal processing on graphs," in *IEEE T-SP*, 2013.
79. K. He, G. Gkioxari, P. Dollar, and R. Girshick, "Mask R-CNN," in *Proc. ICCV*, pp. 2961–2969, 2017.
80. Y. Wang, P.-M. Jodoin, F. Porikli, J. Konrad, Y. Benezeth, and P. Ishwar, "CDnet 2014: An expanded change detection benchmark dataset," in *IEEE CVPRW 2014*, pp. 387–394, 2014.
81. T. Kipf, et al., "Semi-supervised classification with graph convolutional networks," in *ICLR*, 2017.
82. L. Zhao, and L. Akoglu, "PairNorm: Tackling oversmoothing in GNNs," in *ICLR 2020*, 2020.
83. M. Fey, and J. Lenssen, "Fast graph representation learning with PyTorch Geometric," in *ICLRW 2019*, 2019.
84. M. Tezcan, P. Ishwar, and J. Konrad, "BSUV-Net 2.0: Spatio-temporal data augmentations for video-agnostic supervised background subtraction," *IEEE Access*, 9, 53849–53860, 2021.
85. D. Sakkos, H. Liu, J. Han, and L. Shao, "End-to-end video background subtraction with 3D convolutional neural networks," *Multimedia Tools and Applications*, 77, 23023–23041, 2018. https://doi.org/10.1007/s11042-017-5460-9
86. Z. Hu, T. Turki, N. Phan, and J. Wang, "3D atrous convolutional long short-term memory network for background subtraction," *IEEE Access*, 6, 43450–43459, 2018. https://doi.org/10.1109/ACCESS.2018.2861223. keywords: {Three-dimensional displays;Convolution;Machine learning;Kernel;Video sequences;Spatial resolution;Convolutional neural networks;Background subtraction;foreground segmentation;deep learning;3D atrous convolution;convolutional LSTM network}.
87. Y. Wang, Z. Yu, and L. Zhu, "Foreground Detection with Deeply Learned Multi-scale Spatial-Temporal Features," *MDPI Sensors*, 18(12), 4269, 2018. https://doi.org/10.3390/s18124269
88. Lavanya Sharma, A. Singh, D. K. Yadav, "Fisher's Linear Discriminant Ratio based Threshold for Moving Human Detection in Thermal Video", *Infrared Physics and Technology*, **Elsevier,** March, 2016, vo.78, pp. 118–128.

11 Toward Graph Refactoring in Latent Space for Underwater Moving Object Detection for Digital Twins

Meghna Kapoor, Badri Narayan Subudhi, Ankur Bansal, and Thierry Bouwmans

11.1 INTRODUCTION

Moving object detection in a video scene constitutes a fundamental challenge in computer vision. Although numerous surveillance techniques have been developed for outdoor settings, only a few were specifically tailored for underwater applications before the early 2000s. Underwater object detection is primarily used for tasks such as tracking marine life to estimate disease spread [1], detecting cracks in oil and gas pipelines [2], and identifying drowning incidents [3]. Effective sea exploration relies on autonomous vehicles to model the underwater environment using passive camera sensors. Digital twin technology enables the virtual modeling of underwater environments by capturing real-time data through passive camera sensors mounted on autonomous underwater vehicles (AUVs). The data is sent to a system that creates a digital twin, enabling object detection based on the inputs from the sensors, as shown in Figure 11.1. This approach allows for more accurate monitoring and detection in challenging underwater conditions.

Detecting moving objects in underwater environments presents greater challenges compared to traditional in-air scenes because of the inherent characteristics of underwater scenarios. Two principal elements influence underwater images: light absorption and scattering by suspended particles, which produce a hazy effect that diminishes visibility and contrast. Additionally, the water's salinity contributes to optical light attenuation, leading to color distortion due to varying viscosities [4]. These conditions, combined with reduced visibility and color loss, make it difficult for standard computer vision methods to accurately process underwater images. Although numerous conventional object detection algorithms [5, 6] have

DOI: 10.1201/9781003582489-11

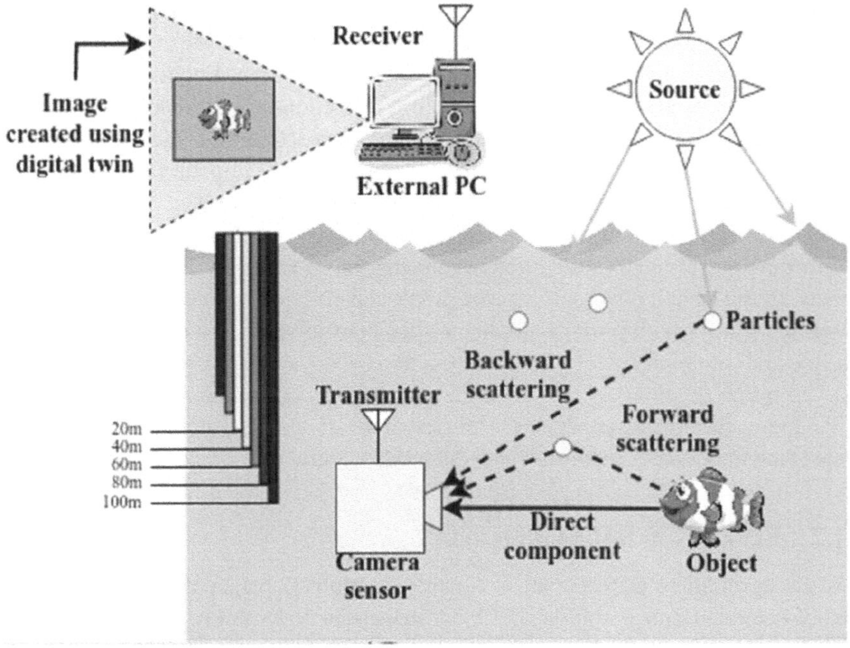

FIGURE 11.1 Using digital twin technology, a virtual model of an underwater object is created with a passive camera sensor. The captured light includes three components: direct light from the object, backward scattered light from particles, and forward scattered light traveling from the object to the camera

been adapted for underwater surveillance, a few have been specifically designed to address these unique underwater challenges, such as haze, color distortion, low visibility, and color loss.

Graph networks have proven more effective in preserving information regardless of data topology and have gained popularity in domains such as action recognition [7], recommendation system [8], and cancer detection [9]. However, the spatial relationships extracted by convolutional networks remain critical for image-processing tasks like object detection. Additionally, CNNs are generally less computationally expensive than graph networks. Thus, combining CNNs with graph networks offers a promising approach to preserving features more effectively for underwater moving object detection.

The unique challenges of detecting moving objects underwater involve extracting features from sequences of underwater images and transforming RGB color images into binary classifications, distinguishing objects from the background. While recent studies [10–13] have made progress in underwater surveillance using various object features, these methods still face significant challenges due to environmental degradations, such as the loss of color and texture information. These degradations undermine the reliability of traditional feature-based approaches. Deep learning techniques [14, 15] have been utilized to separate foreground from background, but they often struggle to retain the fine details of objects. This issue arises partly from noise amplification when applying a spatial kernel in the non-Euclidean domain and additional noise introduced by max pooling operations. Additionally, CNN-based methods face limitations due to non-invertible projections introduced by pooling layers, as well as restrictive symmetry assumptions that impede the extraction of spatial-contextual relationships between pixels. These constraints underscore the need to restructure latent space variables to enhance the capture of node relationships, thus preserving detailed object information in complex underwater environments. In this chapter, we will explore graph construction and refactoring strategies to preserve information more reliably within conventional CNN architectures. For clarity and simplicity, we will focus on the U-Net architecture and demonstrate how to refine relationships in the latent space by mapping them to a graph space. This approach aims to enhance the preservation of spatial-contextual information, improving underwater moving object detection accuracy.

11.2 RESEARCH BACKGROUND

The background of this chapter is organized into two primary areas: underwater object detection and graph-based object detection. Both fields are relatively new, with limited research focused on underwater object detection, while graph theory is still an emerging area of exploration in object detection.

11.2.1 Underwater Object Detection

Underwater object detection can be categorized into two primary approaches: traditional methods and deep learning-based techniques. Traditionally methods often rely on manually produced features and standard algorithms, while deep learning techniques utilize sophisticated neural networks to improve detection precision in difficult underwater environments.

Traditionally, moving objects are identified by analyzing changes in the neighboring regions across consecutive frames. A simple approach involves subtracting subsequent frames to detect local changes [11, 16]. However, these strategies are often ineffective when there are sudden changes in the motion or appearance of objects. Additionally, dynamic backgrounds caused by elements like algae and seaweed further complicate moving object detection. A common method to handle dynamic backgrounds is to model the background using a Gaussian Mixture Model (GMM) [10, 17–21]. Further, some authors have proposed methods relying on the object's color and texture features to identify it [12, 13]. However, color information is often unreliable in underwater environments due to image degradation. Palazzo et al. [22] introduced a method for foreground and background modeling using kernel density estimation to create probabilistic maps, which are then used to detect motion via a Markov field approach. Nonetheless, traditional methods relying on handcrafted features struggle with accurate object detection in underwater scenarios.

Deep learning algorithms extract statistical patterns from training data, utilizing convolution-based designs to exploit spatial correlations in images for object detection [23–25]. To address the limitations of convolutional networks in underwater scenarios, Hua et al. [26] introduced dynamic aggregation and feature enhancement methods for more effective object detection in underwater images. Additionally, transformer-based architectures [27, 28], which leverage attention mechanisms for object detection, often fail to retain shape and structural information critical in complex underwater environments. To mitigate these shortcomings, Bajpai et al. [14] proposed a U-Net architecture with an encoder-decoder structure specifically designed for moving object detection. However, this approach still struggles with the detection of small objects, primarily due to information loss caused by convolution kernels and pooling operations. To overcome these challenges, graph refactoring in latent space offers a promising solution by restructuring node information and preserving spatial-contextual relationships more effectively.

11.2.2 Graph-Based Object Detection

Graph convolutional neural networks (GCNNs) have become increasingly effective in computer vision, excelling in tasks like semantic segmentation and image classification. Xu et al. [29] proposed a novel approach to graph learning, which uses learnable spatial Gaussian kernels to derive contextual insights from sparse graph structures, improving spatial awareness in the process. Similarly, the method by Chen et al. [30] leverages semantic segmentation to draw features from both labels and images, using these features as a foundation for initializing graph connections. GCNNs are especially useful for analyzing data with irregular structures typical of non-Euclidean spaces [31]. For moving object detection, Giraldo et al. [32] developed a GCNN-based approach aimed at detecting objects in complex scenes within previously unseen videos. Their model employs mask R-CNN in addition to features like motion, texture, and color to establish the initial graph. However, this model faces limitations due to its reliance on manually selected features, and current leading techniques often require significant computational resources.

11.3 GRAPH REFACTORING IN LATENT SPACE

In this section, we will explore the concept of graph refactoring in latent space, focusing on techniques to sample neighborhoods and aggregate information for effective message passing between nodes. Graph refactoring in latent space involves restructuring the graph representation to enhance the learning process, especially in the context of underwater moving object detection. By sampling the neighborhood, we can identify relevant nodes that contribute to the detection task, and through aggregation, we combine the information from these nodes to create a more informative and robust representation. This process facilitates efficient message passing, enabling the model to capture complex relationships and dependencies among nodes, which is crucial for accurate object detection in challenging underwater environments.

11.3.1 Motivation and Conceptual Framework

Underwater moving object detection poses unique challenges due to the distinct properties of water. As light waves travel through water, they progressively lose energy, leading to the attenuation of different color channels at varying depths. The red channel is the first to be completely absorbed at approximately 20 feet, causing images to appear predominantly bluish-green at moderate depths. At around 50 feet, the orange color becomes non-existent, followed by the yellow channel at 100 feet. Beyond this depth, the green channel also diminishes, leaving images captured by digital twin technology primarily blue.

Additionally, suspended particles in the water create haze, making it difficult to distinguish the background from the foreground. This can lead to artifacts in the scene, and the interaction of light rays with these particles results in both backward and forward scattering, as seen in Figure 11.1. Forward scattering blurs the image features due to the floating particles. Further, the light rays reach the camera without interacting with the objects causing backward scattering and limiting the contrast of the image. Moreover, the movement of aquatic life is not confined to a two-dimensional plane but can occur in three dimensions, expanding the search space for surveillance systems. These factors make exploring underwater environments far more complex than conventional surveillance systems.

The application of graph theory in moving object detection often relies on features extracted using methods that describe object attributes such as color and texture. However, these attributes are not reliable in underwater environments due to the significant loss of color and texture information. Consequently, traditional graph-based approaches cannot be directly applied. Moreover, computing graphs directly from underwater images can be computationally intensive, making it impractical for real-time applications. Therefore, alternative strategies are needed to effectively leverage graph theory for underwater object detection, focusing on more efficient and robust representations that can accommodate the unique challenges of the underwater environment. This chapter illustrates an interesting problem of using graphs in latent space to refactor the information shared among the latent space variables.

11.3.2 Integration with Deep Learning Models

Graph theory can be effectively integrated with deep learning models by initializing a graph based on the extracted deep features. If these deep features successfully retain coarse details of the image, the graph can provide refinement by incorporating finer details, enhancing the overall quality and accuracy of the detection. This chapter utilizes a deep U-Net architecture that features a pretrained ResNet-50 encoder for the purpose of extracting features. The ResNet-50 network employs convolutional neural networks (CNNs) across multiple levels to derive characteristics from a set of images and map them into a higher-dimensional space. This advanced CNN is particularly effective at identifying spatial relationships between pixels, assuming that the underlying data is structured in Euclidean space. However, while convolution operations are adept at converting information from a lower-dimensional image space to a more complex latent space, they often struggle to maintain the spatial and contextual fidelity of the original underwater images in this enhanced representation. To overcome this challenge, it is crucial to develop a feature pooling or reweighting module that can strengthen and clarify the connections within the latent space.

Once the features have been re-weighted, they are mapped back to the original space using an inverse-mapping function implemented by the decoder. To maximize the retention of information throughout this process, the U-Net architecture employs an identical mapping strategy. The skip connections in U-Net between encoder and decoder layers are pivotal, as they directly transfer features from corresponding encoder layers to decoder layers. This mechanism helps preserve high-resolution details that could otherwise be lost during the down-sampling and up-sampling stages of the network.

To further boost the model's performance, we initialize the model with pretrained weights from ImageNet, taking advantage of transfer learning. The model is then fine-tuned using the F4Knowledge dataset, allowing the weights to be updated and better adapted to the specific characteristics and challenges posed by underwater moving object detection. This strategy enables the model to leverage a strong foundational understanding while specializing in the nuances of the target domain, thereby improving detection accuracy and robustness.

11.3.3 Graph Construction

A graph G can be defined over the non-Euclidean latent space as an unordered set of tuples. This set comprises edges (E) and nodes or vertices (V). In this structure, the nodes or vertices are connected by links or edges. Features such as color, motion, or texture can be extracted and tailored to specific attributes, as demonstrated by Giraldo et al. [33]. However, defining these attributes in the underwater domain is challenging due to environmental complexities. To enhance performance, higher-order space features can also be considered. In this chapter, we explore approaches that initialize the graph by extracting features from the latent space using a deep learning architecture. Each element of the latent vector is assumed to represent a node in the graph.

11.3.3.1 Sampling

Neighborhood N in graph is defined based on the direct connections or hops along a path, as illustrated in Figure 11.2(a). It represents a fixed subset drawn uniformly from the sample set, with neighbors being updated at each iteration. Utilizing neighbors aids in minimizing both computational complexity and resource requirements. The information that is acquired from such neighbors is aggregated and provided to the selected node in the subsequent phase or layer.

11.3.3.2 Aggregator Functions

The information among the nodes are aggregated using an aggregator function a_f that delineates the node relationship. The node information within a specific neighborhood is disseminated, and the overall output is calculated utilizing an aggregator function as shown in Figure 11.2(b). In latent space layer (l), neighbors (j) are represented by h_j^{l-1} which lacks a specific order, and the aggregation of the individual node remains trainable.

$$h_{\mathcal{N}_i}^l \leftarrow a\_f\left\{h_j^{l-1} \quad \forall j \in \mathcal{N}(i)\right\} \tag{11.1}$$

11.3.4 GraphSage

GraphSage stands for Graph sampling and aggregation. Employing inductive reasoning GraphSage is utilized for the generation of large graphs. The basic architecture of the GraphSage algorithm is given in Figure 11.2.

Further, different stages of the GraphSage are given as follows. In this model, three different aggregator functions are used to re-weight the latent space relationship.

Mean Aggregator: The mean of neighborhood nodes is taken into account to evaluate the information at the current node.

Max Aggregator: The max or the pooling function operates by doing element-wise max across the neighboring nodes.

LSTM Aggregator: Compared to the above two functions, the third is the most complex function and is inherently not symmetric. Random permutation among the neighbors is applied.

11.3.4.1 Refactoring Using GraphSage

Every element of the feature space is considered as a node. The information from neighboring nodes N is defined over the information from previous nodes. An

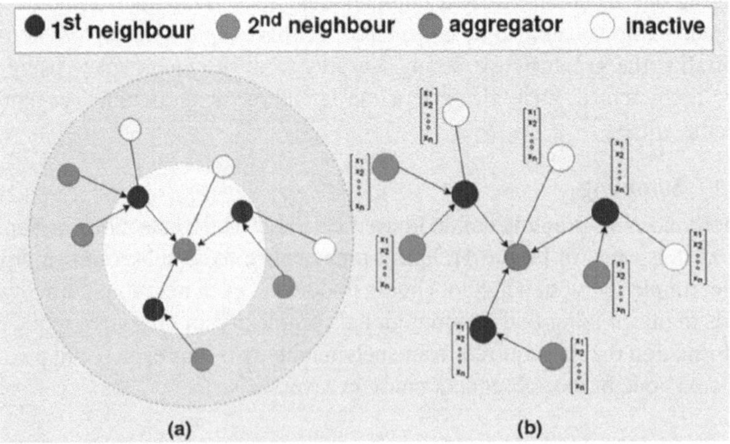

FIGURE 11.2 Graph algorithm. (a) Node sampling: nodes are sampled to define a neighborhood based on hop distance, (b) aggregation of message: the message among neighbors is aggregated to update the refactored message

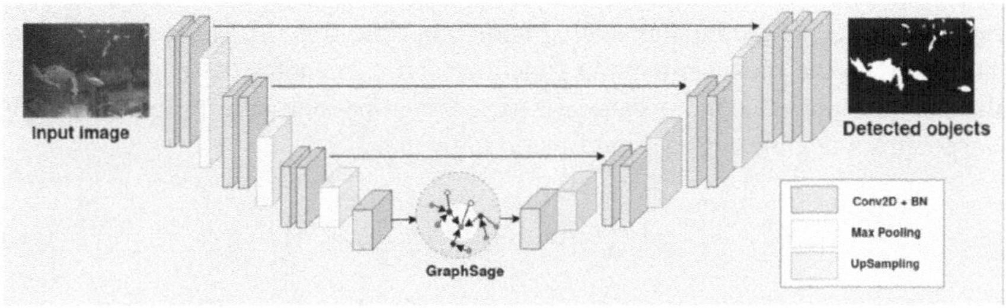

FIGURE 11.3 Moving object detection using GraphSAGE-based refactoring: The input image is passed through an encoder network with Conv2D layers, batch normalization, and max pooling to extract feature representations as a latent vector. This vector initializes the GraphSAGE-based refactoring, which is then processed by a decoder with upsampling and Conv2D layers

aggregator function from the set {mean, max, LSTM} is applied over the obtained information from neighbors and is denoted as h^{l}_{N}. The current information and information from the neighborhood are concatenated. The obtained vector is passed through a non-linear activation (sigmoid in our case). The updated representation of node i in layer l is given as:

$$h_i^l = f_{\text{update}}\left(h_{\mathcal{N}_i}^l, h_j^{l-1}\right). \tag{11.2}$$

Here, f_{update} can be simply an aggregator operator or any complex function. We have used the update function as a concatenate operator.

11.3.5 Refactoring Using Principal Neighborhood Aggregation

Xu et al. [34] demonstrated that mean and max aggregators are insufficient for distinguishing features with identical characteristics but different cardinalities. The graph structure information cannot be effectively captured by the mean and max operators used by [35] to refactor the latent vector accurately as shown in 11.3.

To address this limitation, Corso et al. [36] proposed a principal neighborhood aggregation approach that employs multiple aggregators and scalers to combine information from neighboring nodes, which can be used to refactor the feature vector in high-dimensional space. Hence, a combination of multiple aggregators can be tested for underwater moving object detection.

The graph learning rule, according to the message-passing neural network (MPNN) framework, is defined as:

$$h_v^{t+1} = U_t\left(h_v^t, \sum_{\omega \in \mathcal{N}(v)} M_t\left(h_v^t, h_w^t, e_{vw}\right)\right) \tag{11.3}$$

where h_v^t represents the hidden state at node v at time step t, $U_t$is the vertex update function, M_t is the message-passing function, and e_{vw} denotes the edge feature. To enhance this model, a more generalized message-passing rule proposed by [36] incorporates a flexible aggregation function, given by:

$$h_v^{t+1} = U_t\left(h_v^t, \oplus_{\omega\in\mathcal{N}(v)} M_t\left(h_v^t, h_w^t, e_{vw}\right)\right) \tag{11.4}$$

Here, $\oplus$ is a generalized aggregation operator defined as:

$$\oplus = \begin{bmatrix} I \\ A \\ D \end{bmatrix} \otimes \begin{bmatrix} max_{j\in\mathcal{N}_i} X_j^l \\ mean_{j\in\mathcal{N}_i} X_j^l \\ \sigma^2(X) \\ \sigma(X) \end{bmatrix} \tag{11.5}$$

The scalar values in the first matrix are used to either amplify or attenuate the messages in a multi-layer network. A minor variation can exponentially affect the gradients, making it crucial to use multiple scalars combined with a generalized aggregation operator. In the expression, I represents the identity function, A denotes amplification, and D signifies attenuation. The second matrix is a set of aggregation functions: max, mean, variance (σ^2), and standard deviation (σ).

Max and mean aggregation functions are straightforward, defined as:

$$\max_i\left(X^l\right) = \max_{j\in\mathcal{N}_i} X_j^l, \tag{11.6}$$

$$\text{mean}_i\left(X^l\right) = \text{mean}_{j\in\mathcal{N}_i} X_j^l \tag{11.7}$$

Variance $\sigma^2(X)$ and standard deviation $\sigma(X)$ are utilized to diversify the information at each node, expressed in terms of expectation E as:

$$\sigma(X) = \sqrt{E\left[X^2\right] - \left(E[X]\right)^2} \tag{11.8}$$

$$\sigma^2(X) = E\left[X^2\right] - \left(E[X]\right)^2. \tag{11.9}$$

The degree of a node, d, is used to proportionally scale the aggregated information. Even a slight change in the degree can lead to significant attenuation or amplification.

Thus, a scaling function $S(d,\alpha)$ is applied to the aggregated tensor:

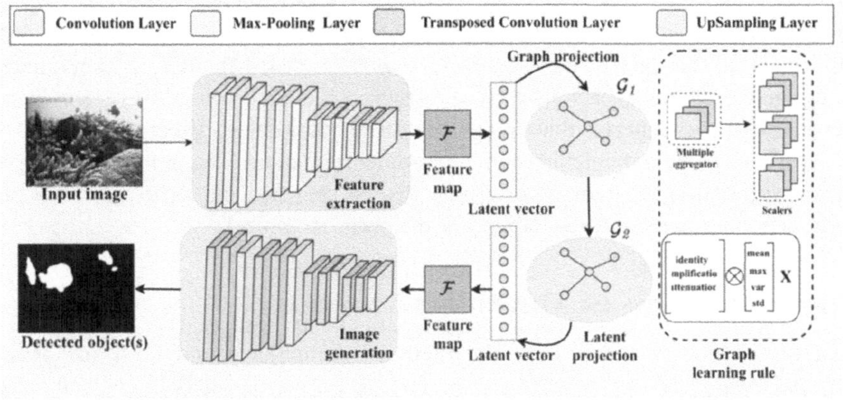

FIGURE 11.4 End-to-end architecture of PNA-based graph refactoring: the input image is processed by a feature extraction module to generate a feature representation. The output from the last layer is flattened into a latent vector, which initializes a graph. Multiple scalers and aggregators are used to refactor the graph messages. After two graph layers, the latent vector is employed by the image generation module to detect objects

$$S(d,\alpha)=\left(\frac{\log(d+1)}{\delta}\right)^{\alpha}, \tag{11.10}$$

where $\alpha = 0$ for identity function (I), $\alpha = -1$ for attenuation (D), and $\alpha = +1$ for amplification (A).

Two principal neighborhood graph learning layers are utilized to restructure node relationships and facilitate message passing among the nodes as shown in Figure 11.4.

11.4 EXPERIMENTAL SETUP AND RESULTS

11.4.1 Implementation Details

We have used NVIDIA A100 80 GB GPU to execute the proposed methodology. The device has 128 GB of RAM, and the method is developed using Python and the PyTorch framework on a Linux OS. We assessed the efficacy of the suggested strategy utilizing the F4Knowledge benchmark database. During training, a batch size of 2 was employed. For the GraphSAGE and PNA implementations, we considered two hops of neighbors to effectively aggregate information from surrounding nodes. Additionally, Adam optimizer is used in proposed method with a learning rate of 10^{-3} to facilitate efficient convergence. The U-Net architecture utilized binary cross-entropy as the loss function for hyperparameter updates and gradient computation.

To evaluate the model's performance, we conducted tests using various aggregation functions and analyzed the results both visually and quantitatively. The effectiveness of the proposed methods was validated against 11 SOTA techniques: MLCB [37], Texture-BGS [38], MLBGS [39], SubsenseBGS [40], SILTP [41], MFI [12], GSMM [17], AGMM [20], ABMM [21], ADE [42], and GWFT [18].

Fish4Knowledge (F4Knowledge) consists of video sequences recorded by ten distinct cameras in aquatic settings. Our research concentrated on five distinct difficulties within this dataset: complex backgrounds, crowded challenges, camouflaged foregrounds, dynamic backgrounds, and hybrid scenes. Each of these challenges presents distinct scenarios that complicate underwater object detection due to varying levels of visual complexity and environmental factors. The number of samples differs across these challenges, offering a diverse set of conditions to assess the performance robustness and adaptability of the proposed model.

11.4.2 Qualitative and Quantitative Analysis

The visual assessment of both the methods: GraphSage and PNA for detecting underwater moving objects was performed on various challenging sequences from the Fish4Knowledge database and the Underwater Change Detection dataset. Figure 11.5 illustrates the results for the Fish4Knowledge dataset. Columns (a) and (b) present the original images and the ground truth of the sequences, respectively. Columns (c) to (h) of Figure 11.5 display the results obtained from several techniques, including Texture-BGS [38], MLBGS [39], MLCB [37], SubsenseBGS [40], SILTP [41], and MFI [12]. It is evident that many state-of-the-art methods struggle to capture the entire region of the objects, often missing the moving object areas entirely in several instances. In contrast, the results produced by GraphSage [35] and PNA [43], as shown in Figure 11.5 columns (i) and (j), respectively, successfully detect the objects accurately, demonstrating their effectiveness in challenging underwater conditions.

An investigation is performed to assess the quantitative performance of the proposed system utilizing metrics like accuracy, recall, precision, and F-measure. Table 11.1 shows the average F-measure obtained from nine SOTA methods. The proposed method is compared against 13 state-of-the-art methods: GMM [19], KDE [44], SuBSENSE [45], Vu Meter [46], Wronksian [10], ML-BGS [39], MC-BGS [37], FgSegNet [47], MFI [12], CSLTP [13], MSGAN [15], HMLS [48],

(a) (b) (c) (d) (e) (f) (g) (h) (i) (j)

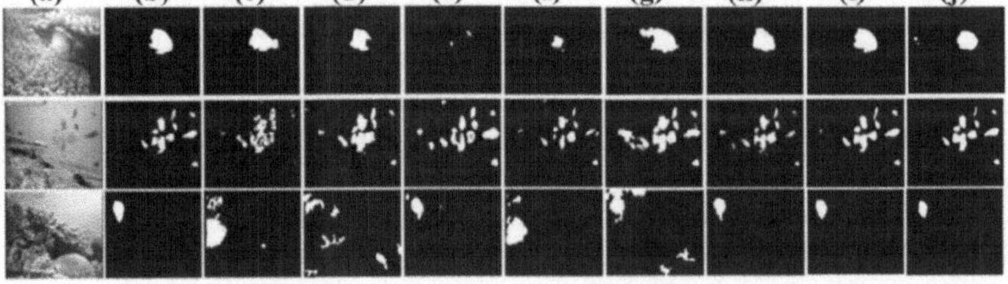

FIGURE 11.5 Visual analysis of the proposed method with SOTA methods on Fish4Knowledge dataset. (a) Original image, (b) ground-truth, (c) Texture-BGS [38], (d) ML-BGS [39], (e) MultiCueBGS [37], (f) SuBSENSEBGS [45], (g) SILTP [41], (h) MFI [12], (i) GraphSage, (j) PNA. (a) Aggregated quantitative measure of (b) Comparison of different aggregator function F4Knowledge dataset on proposed model

TABLE 11.1
Average F-Measure on Four Challenges of Fish4Knowledge Dataset

Method	ComplexBkg	DynamicBkg	Crowded	Hybrid	Overall
GMM [19]	0.12	0.06	0.26	0.15	0.15
ML-BGS [39]	0.58	0.32	0.74	0.46	0.52
MC-BGS [37]	0.48	0.33	0.68	0.72	0.55
KDE [44]	0.13	0.07	0.20	0.16	0.14
SuBSENSEBGS [45]	0.21	0.81	0.67	0.42	0.53
FgSegNet [47]	0.64	0.39	0.68	0.60	0.58
MFI [12]	0.83	0.64	0.69	0.64	0.70
CSLTP [13]	0.80	0.76	–	0.83	0.79
MSGAN [15]	0.92	0.79	0.74	0.89	0.83
HMLS [48]	0.84	0.90	0.84	0.91	0.87
GraphSage [35]	0.99	0.99	0.98	0.98	0.99
PNA [43]	0.99	0.98	0.99	0.99	0.99

and GraphSage [35]. The proposed method outperforms the current state-of-the-art methods, showing a significant improvement of around 20% in the overall F-measure on the Fish4Knowledge dataset [49] compared to the learning-based method, MSGAN [15].

While GraphSage [35] utilizes a single aggregator across the neighborhood, PNA employs four different aggregators to enhance message passing. This multi-aggregator approach allows for better performance in crowded and hybrid environments, making it particularly effective in real-life underwater scenarios with multiple fish present. Overall, our method outperforms all existing methods in terms of detection accuracy and robustness in complex underwater environments.

11.4.3 Ablation Study

An ablation study is performed to assess the proposed scheme by applying different aggregation methods, such as max, mean, and LSTM. The ablation study is performed on the F4Knowledge dataset. The results are detailed in Table 11.2,

TABLE 11.2
Ablation Study of Different Aggregation Functions in Terms of Accuracy, Precision, Recall, and F-Measure on the F4Knowledge Dataset

Function	Accuracy	Precision	Recall	F-measure
Mean	0.9851	0.9938	0.9115	0.9866
Max	0.9446	0.9901	0.9872	0.9866
LSTM	0.9850	0.9919	0.9389	0.9868

and a comparison of the various aggregators is illustrated in Figure 11.6. The study shows that the mean operator achieves the highest accuracy and precision, with an F-measure only 0.02 lower than the best-performing method. Additionally, the mean operator is computationally more efficient compared to LSTM, making it a preferable choice due to its lower computation time.

Table 11.3 presents the results of an ablation study that explores the impact of various aggregators and scalers on the performance of the proposed algorithm in complex background scenarios, measured in terms of the F-measure. The study reveals that the choice of different scalers and aggregators significantly influences the algorithm's effectiveness. Specifically, certain aggregators may overlook information from some nodes, leading to the introduction of noise in the process. Therefore, selecting appropriate aggregators and scalers is crucial for optimizing performance. The table includes the mean, max, standard deviation (std), and variance (var) as the aggregators, along with identity, amplification, and aggregation as the scalers that retain the most information. This suggests that the combination of these specific scalers and aggregators is particularly effective in maintaining the integrity of the data throughout the processing pipeline.

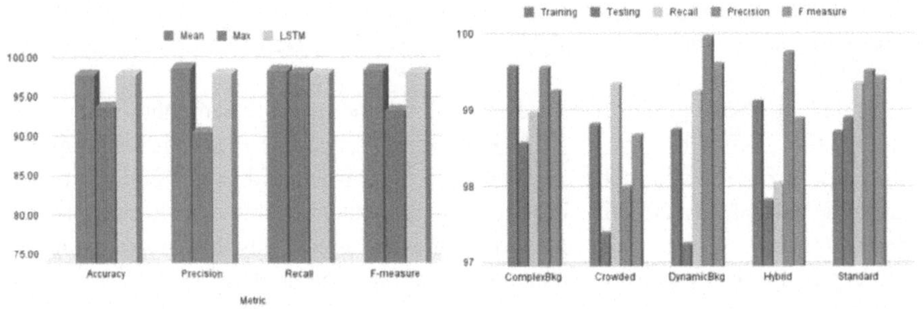

FIGURE 11.6 Quantitative performance of Fish 4 Knowledge dataset using the GraphSage

TABLE 11.3
Ablation Study Performed on Complex Background Challenge of Fish4Knowledge Dataset with a Different set of Aggregators and Scalers

Aggregator	Scalers	F-Measure
Mean	Identity, Amplification	99.26
Mean	Identity, Amplification, attenuation	99.30
Max	Identity, Amplification	99.26
Max, std	Identity, Amplification	99.41
Mean, std, var	Identity, Amplification	99.13
Mean, max, std, var	Identity, Amplification, attenuation	99.46

11.5 APPLICATIONS AND IMPLICATIONS

Detecting moving objects in underwater scenarios opens up numerous applications, especially when integrated with autonomous underwater vehicles (AUVs) using digital twin technology. For instance, by monitoring aquatic life, we can apply disease control measures more effectively. Additionally, in the case of damage to oil and gas pipelines, moving object detection algorithms can help identify potential leaks by detecting bubbles. Furthermore, in fish farming, these algorithms can accurately measure the size and shape of objects, providing valuable insights for managing fish populations and optimizing growth conditions. The algorithms discussed in this chapter provide a foundation for these applications, offering robust methods for analyzing underwater environments.

11.6 CONCLUSIONS

This chapter demonstrates the application of digital twin technology for detecting moving objects in complex underwater environments, utilizing graph-based techniques to enhance detection accuracy by addressing noise introduced during latent space projections. Sampling and aggregation operations within the graph space are employed to refine information transfer among nodes. However, the mean and max operations do not account for node cardinality, leaving it unclear how many nodes contribute to the message. To address this, principal graph aggregation is applied, using multiple aggregators and scalers to mitigate the effects of node cardinality, thereby improving information propagation and detection performance.

Currently, no established method exists to determine how latent space nodes should be connected. We randomly initialize the connections in this approach, but exploring alternative initialization distributions could provide valuable insights and potentially enhance the model's performance. In future, methods like K-Nearest Neighbors (KNN) and distance-based initialization using cosine similarity could be explored to establish more meaningful connections. Additionally, different graphs can be constructed for various modalities – such as motion data and RGB images – ensuring that corresponding nodes across graphs represent the same object. By fusing information within the graph space, we aim to leverage complementary data from each modality to improve the robustness and accuracy of moving object detection in underwater scenes.

REFERENCES

1. Daoliang Li and Ling Du. "Recent advances of deep learning algorithms for aquacultural machine vision systems with emphasis on fish." *Artificial Intelligence Review* (2022)1–40.
2. Juhyun Kim et al. "The development of leak detection model in subsea gas pipeline using machine learning." *Journal of Natural Gas Science and Engineering* 94 (2021), p. 104134.
3. Fei Lei et al. "Drowning behavior detection in swimming pool based on deep learning." *Signal, Image and Video Processing* (2022)1–8.

4. Deepak Kumar Rout et al. "Underwater visual surveillance: A comprehensive survey." *Ocean Engineering* 309 (2024), p. 118367.
5. Thierry Bouwmans. "Traditional and recent approaches in background modeling for foreground detection: An overview." *Computer Science Review* 11 (2014), pp. 31–66.
6. Marie-Neige Chapel and Thierry Bouwmans. "Moving objects detection with a moving camera: A comprehensive review." *Computer Science Review* 38 (2020), p. 100310.
7. Himanshu Singh et al. "Action recognition in dark videos using spatiotemporal features and bidirectional encoder representations from transformers." *IEEE Transactions on Artificial Intelligence* 4.6 (2022), pp. 1461–1471.
8. Tian Xie et al. "Graph dynamical networks for unsupervised learning of atomic scale dynamics in materials." *Nature Communications* 10.1 (2019), p. 2667.
9. Nazar Zaki, Wenjian Qin, and Anusuya Krishnan. "Graph-based methods for cervical cancer segmentation: Advancements, limitations, and future directions." *AI Open* 4 (2023), pp. 42–55.
10. Deepak Kumar Rout et al. "Spatio-contextual Gaussian mixture model for local change detection in underwater video." *Expert Systems with Applications* 97 (2018), pp. 117–136.
11. Deepak Kumar Rout et al. "A novel five-frame difference scheme for local change-detectioninunderwatervideo." In *Proceeding of Fourth International Conference on Image Information Processing (ICIIP)*. IEEE. 2017, pp. 1–6.
12. Srikanth Vasamsetti et al. "Automatic underwater moving object detection using multi-feature integration framework in complex backgrounds." *IET Computer Vision* 12.6 (2018), pp. 770–778.
13. Srikanth Vasamsetti et al. "3D local spatio-temporal ternary patterns for movingobject-detectionincomplexscenes." *Cognitive Computation* 11 (2019), pp. 18–30.
14. Vatsalya Bajpai et al. "Underwater U-net: Deep learning with U-net for visual underwater moving object detection." In *Proceedings of OCEANS 2021: San Diego–Porto*. 2021, pp. 1–4.
15. Prashant W. Patil et al. "Motion saliency based generative adversarial network for underwater moving object segmentation." In *Proceedings of the IEEE International Conference on Image Processing (ICIP)*. 2019, pp. 1565–1569.
16. M. R. Prabowo et al. "A moving objects detection in underwater video using subtraction of the background model." In *2017 4th International Conference on Electrical Engineering, Computer Science and Informatics (EECSI)*. 2017, pp. 1–4. doi: 10.1109/EECSI.2017.8239148.
17. Martin Radolko and Enrico Gutzeit. "Video segmentation via a Gaussian switch background model and higher order Markov Random fields." In *VISAPP (1)*. 2015, pp. 537–544.
18. Martin Radolko, Fahimeh Farhadifard, and Uwe von Lukas. "Change detection in crowded underwater scenes-via an extended gaussian switch model combined with a flux tensor pre-segmentation." In *International Conference on Computer Vision Theory and Applications*. Vol. 5. Scitepress. 2017, pp. 405–415.
19. Chris Stauffer and W. Eric L. Grimson. "Learning patterns of activity using real-time tracking." *IEEE Transactions on Pattern Analysis and Machine Intelligence* 22.8 (2000), pp. 747–757.
20. Zoran Zivkovic. "Improved adaptive Gaussian mixture model for background subtraction." In *Proceedings of the 17th International Conference on Pattern Recognition, 2004. ICPR 2004*. Vol. 2. IEEE. 2004, pp. 28–31.
21. Lavanya Sharma, A. Singh, D. K. Yadav, "Fisher's Linear Discriminant Ratio based Threshold for Moving Human Detection in Thermal Video", *Infrared Physics and Technology*, Elsevier, March, 2016, vol 78, pp. 118–128

22. S. Palazzo, I. Kavasidis, and C. Spampinato. "Covariance based modeling of under waterscenes for fish detection." In *2013 IEEE International Conference on Image Processing*. 2013, pp. 1481–1485. doi: 10.1109/ICIP.2013.6738304.
23. Xiu Li et al. "Fast accurate fish detection and recognition of underwater images with fast r-cnn." In *OCEANS 2015-MTS/IEEE Washington*. IEEE. 2015, pp. 1–5.
24. Hong Liu, Pinhao Song, and Runwei Ding. "Towards domain generalization in underwater object detection." In *2020 IEEE International Conference on Image Processing (ICIP)*. IEEE. 2020, pp. 1971–1975.
25. Xutao Liang and Pinhao Song. "Excavating roi attention for underwater object detection." In *2022 IEEE International Conference on Image Processing (ICIP)*. IEEE. 2022, pp. 2651–2655.
26. Xia Hua et al. "Underwater object detection algorithm based on feature enhancement and progressive dynamic aggregation strategy." *Pattern Recognition* 139 (2023), p. 109511.
27. Yang Liu et al. "DP-FishNet: Dual-path pyramid vision transformer-based underwater fish detection network." *Expert Systems with Applications* 238 (2024), p. 122018.
28. Gangqi Chen et al. "HTDet: A Hybrid Transformer-Based Approach for Underwater Small Object Detection." *Remote Sensing* 15.4 (2023), p. 1076.
29. Hang Xu et al. "Spatial-aware graph relation network for large-scale object detection." In *Proceedings of the IEEE/CVF Conference on Computer Vision and Pattern Recognition*. 2019, pp. 9298–9307.
30. Shengjia Chen, Zhixin Li, and Zhenjun Tang. "Relation R-CNN: A graph based relation-aware network for object detection." *IEEE Signal Processing Letters* 27 (2020), pp. 1680–1684.
31. Zonghan Wu et al. "A comprehensive survey on graph neural networks." *IEEE Transactions on Neural Networks and Learning Systems* 32.1 (2020), pp. 4–24.
32. Jhony H Giraldo et al. "Graph CNN for moving object detection in complex environments from unseen videos." In *Proceedings of the IEEE/CVF International Conference on Computer Vision*. 2021, pp. 225–233.
33. Jhony H Giraldo, Sajid Javed, and Thierry Bouwmans. "Graph moving object segmentation." *IEEE Transactions on Pattern Analysis and Machine Intelligence* 44.5 (2020), pp. 2485–2503.
34. Keyulu Xuetal. "Howpowerfularegraphneuralnetworks?" *arXivpreprint arXiv:1810. 00826* (2018).
35. Meghna Kapoor et al. "Underwater moving object detection using an endto-end encoder-decoder architecture and graphSage with aggregator and refactoring". In *Proceedings of the IEEE/CVF Conference on Computer Vision and Pattern Recognition*. 2023, pp. 5635–5644.
36. Gabriele Corso et al. "Principal neighbourhood aggregation for graph nets." *Advances in Neural Information Processing Systems* 33 (2020), pp. 13260–13271.
37. Seung Jong Noh and Moongu Jeon. "A new framework for background subtraction using multiple cues." In *Asian Conference on Computer Vision*. Springer. 2013, pp. 493–506.
38. Marko Heikkila and Matti Pietikainen. "A texture-based method for modeling the background and detecting moving objects." *IEEE Transactions on Pattern Analysis and Machine Intelligence* 28.4 (2006), pp. 657–662.
39. Jian Yao and Jean-Marc Odobez. "Multi-layer background subtraction based on color and texture." In: *Proceedings of the IEEE/CVF Conference on Computer Vision and Pattern Recognition*. IEEE. 2007, pp. 1–8.

40. Pierre-Luc St-Charles, Guillaume-Alexandre Bilodeau, and Robert Bergevin. "Flexible background subtraction with self-balanced local sensitivity." In *Proceedings of the IEEE Conference on Computer Vision and Pattern Recognition Workshops*. 2014, pp. 408–413.
41. Hong Han et al. "Moving object detection revisited: Speed and robustness". *IEEE Transactions on Circuits and Systems for Video Technology* 25.6 (2014), pp. 910–921.
42. Zoran Zivkovic and Ferdinand Van Der Heijden. "Efficient adaptive density estimation per image pixel for the task of background subtraction." *Pattern Recognition Letters* 27.7 (2006), pp. 773–780.
43. Meghna Kapoor et al. "Principal graph neighborhood aggregation for underwater moving object detection." In *International Conference on Pattern Recognition (ICPR)*. IEEE. 2024.
44. Ahmed Elgammal et al. "Background and foreground modeling using nonparametric kernel density estimation for visual surveillance." *Proceedings of the IEEE* 90.7 (2002), pp. 1151–1163.
45. Pierre-Luc St-Charles, Guillaume-Alexandre Bilodeau, and Robert Bergevin. "SuBSENSE: A universal change detection method with local adaptive sensitivity." *IEEE Transactions on Image Processing* 24.1 (2014), pp. 359–373.
46. Yann Goyat et al. "Vehicle trajectories evaluation by static video sensors." In: *Proceedings of the IEEE Intelligent Transportation Systems Conference*. 2006, pp. 864–869.
47. Long Ang Lim and Hacer Yalim Keles. "Foreground segmentation using convolutional neural networks for multiscale feature encoding." *Pattern Recognition Letters* 112 (2018), pp. 256–262.
48. Ahmad Salman et al. "Automatic fish detection in underwater videos by a deep neural network-based hybrid motion learning system." *ICES Journal of Marine Science* 77.4 (2020), pp. 1295–1307.
49. Robert B Fisher et al. *Fish4Knowledge: Collecting and analyzing massive coral reef fish video data*. Vol. 104. Springer, 2016.

12 Toward a Complete System for Pupil Monitoring Using Deep Learning for Digital Twins

Kamal Sehairi, Patrick Franco, El-Hadi Zahzah, and Thierry Bouwmans

12.1 INTRODUCTION

The education sector in the world, especially developing countries, is suffering from a severe phenomenon of school dropout, where many students lose the opportunity to complete their education each year, posing a threat to the future of education and society. School dropout statistics vary significantly across the globe. Globally, there are about 16% children and youth (from primary to senior secondary) who are not attending the school. At the primary level, one out of ten children worldwide is not in school. About 122 million, or 48% of the out-of-school population, are young women and girls. In Africa, one out of five children does not attend school (19.7%), and only half of the children attend upper secondary school [1]. Various factors can cause school dropout and truancy, such as school environment, bullying, harassment, discrimination, poor relationships, or home environments, such as parents' divorce, death of parents, financial difficulties, distance from school, and lack of transportation.

With the increasing complexity of this phenomenon, it has become necessary to search for innovative solutions to combat some of these factors. Artificial intelligence and digital twins' technology offer immense potential to combat school dropout by providing effective monitoring and analysis tools and developing intelligent systems that help assess student behavior and provide them with the necessary support. In this context, we propose a vision-based attendance and behavior recognition of pupils and students. This system is implemented on a high-end embedded system and uses IoT technologies for real-time monitoring and tracking. The rest of this chapter is organized as follows: a short overview of techniques, studies, and intelligent systems used for reducing dropout is given in Section 12.2. In Section 12.3, we detail the methodology for developing a comprehensive IoT-based student attendance system, along with an extension for student activity monitoring using twin

DOI: 10.1201/9781003582489-12

technology. Section 12.4 presents the implementation and results. Finally, Section 12.5 summarizes and presents a discussion of potential future work.

12.2 STUDIES, TECHNIQUES, AND INTELLIGENT SYSTEMS FOR DROPOUT REDUCTION

12.2.1 Dropout Studies and Machine Learning Predictors

Many studies, innovative techniques, and advanced systems have been conducted and developed to pinpoint the underlying causes of student dropout and propose effective solutions to reduce it, ensuring students remain engaged and enrolled.

For example, Reschly et al. [2] enumerated several variables that affect the dropout rate. They classified these variables into three factors: (1) student-related aspects such as grade repetition, low academic achievement, attendance, and behavior, (2) family-related elements like parental monitoring and educational support, and (3) school-related factors, including school climate, teacher–student ratio, and discipline policies. Ogresta et al. [3] highlighted the diverse reasons for dropout based on a study conducted in Croatia. They found that the poor academic achievements, the lack of motivation to attend school, the high absenteeism, the bullying, the grade retention, and the parents' irresponsibility are the main causes of the high rate of school dropout. In their technical report, Hoff et al. [4] state that attendance or absenteeism associated with students' behavior and course performance are the key factors of school dropout. Archambault et al. [5] showed that there is a strong connection between student engagement and dropout. They divided the causes of student disengagement that leads to school dropout into three main factors: macro, micro, and individual ones. Macro elements are associated with the global environment (rural/urban) and societal standards (social acceptance, low-socioeconomic status neighborhood, immigration, and minority integration). Micro aspects include the microenvironment, such as family condition and responsibility (parental monitoring, poverty, drought history), school climate, and teacher's attitudes toward students. For more studies about dropout factors and causes, we refer the reader to the references [7–10].

Other studies tried to predict school dropout using machine learning techniques based on various factors such as attendance records, academic performance, student behavior, and student interaction. For example, Psyridou et al. [11] used a dataset collected in 13 years incorporating data from kindergarten to upper secondary education to train four classifiers (Balanced Random Forest, Adaboost Ensemble, Adaboost, and the Bagging Decision Tree). This dataset includes students' academic and cognitive skills, motivation, well-being, behavior, family background, bullying, media usage, and recorded dropout. Results showed that the classifier model achieved a mean area under the curve (AUC) of 0.65, which indicates that the model has some ability to predict school dropout that may help teachers and parents identify at-risk students. Hernândez et al. [12] trained various classifiers (Random Forest, J48 decision tree, and Logistic regression) using WEKA [13] to identify students at risk of dropping out by the end of their first semester. The dataset was collected over three years and consisted of more than 58K records of student grades. A *k*-means algorithm was applied to this dataset to cluster the academic performance of students into three groups (low-performance, intermediate performance, and high-performance). These classes were used to train the classifiers. Statistics showed that Random Forest provides the best result with 80% accuracy. Cuevas-Chávez et al. [14] developed a predictor for school dropout. The authors employed the RobustScalar method as a preprocessing step to scale the Predict Students' Dropout and Academic Success dataset [15]. This dataset includes a wide range of information, such as demographic, macroeconomic, and socioeconomic data, as well as student enrolment details, for 17 undergraduate degrees spanning different fields of study in both the first and second semesters. The dataset has three classes: enrolled, dropout, and graduate. Because this dataset has unbalanced classes, different resampling techniques were employed to generate synthetic data for the minority class (Oversampling using Adaptive synthetic or Synthetic Minority Oversampling Technique with Support Vector Machine) or reduce the number of samples for the majority class (Undersampling + Oversampling using the Synthetic Minority Oversampling Technique with the Edit Nearest Neighbor). Three classifiers were used: Random Forest, Support vector machine, and XGBoost with Bayesian Optimization for hyperparameters configuration. Tests showed that the Synthetic Minority Oversampling Technique with the Edit Nearest Neighbor combined with the Support vector machine classifier achieved the best result with an accuracy of 93.55% and a precision of 94.11%. In [16], Freitas et al. used socioeconomic data to develop an IoT platform for school dropouts. This data contains information on 1549 students recorded from 2008 to 2019. Information includes human development index, age, gender, race by neighborhood, family source of income, higher school education provenance, and distance from the campus. Various machine learning classifiers were trained, including Decision Tree, Logistic Regression (LR), Support Vector Machine (SVM), K-nearest neighbors, Multilayer perceptron, and Deep Learning (DL). The IoT system is responsible for connecting different devices, performing predictions, storing records, and sharing results. Outcomes show that the developed system achieves an accuracy of 99.34%, 99.34% F1 score, 100% recall, and 98.69% precision using Decision Tree (DT). Other dropout predictors can be found in [17–19].

12.2.2 Intelligent Systems for Dropout Prevention

All these studies and machine learning predictors (Section 12.2.1) emphasize the critical role of consistent attendance and active parental monitoring in effectively reducing school dropout rates. While the dropout predictors focus on students' grades and sometimes absenteeism, bullying, and family background, other systems focus on school dropout prevention based on the factors above. These intelligent systems include attendance systems, personalized learning plans, behavior recognition, and continuous tracking. For example, Menezes et al. [20] developed a student attendance system using various image-capturing devices (iPhone 7, Moto G, and webcam). The authors employed the histogram of oriented gradients HOG [21] for face detection and the deep learning FaceNet [22] model for face recognition. Experiments show that the acquisition device (image resolution) considerably affects the results of the detection, where the iPhone 7 achieved 97% accuracy, the Moto G 91.9%, and the webcam 51.2%. Kuang and Baul [23] developed a similar attendance system. They applied the Haar cascade classifier [24] for face detection and the FaceNet DNN model for face recognition. Implementation was done on a laptop with Intel® core™ i5-7200U CPU @2.50GHz, RAM 8.00GB. Test results showed an accuracy of 95%. El Mrabet and Ait Moussa [25] developed an IoT system for school attendance using RFID (Radio Frequency Identification) tags. The system detects and saves the registration document and then sends it to the administration. In addition, the system warns parents via Email or SMS about absenteeism and automatically forwards the missed lessons. Athanesious et al. [26] proposed the Deep Learning-based Automated Attendance System (DPAAS). The system uses a modified Single-Shot MultiBox Detector SSD [27] for face detection and the VGG deep neural network [28] for multi-class face recognition. Training of the face detector was done using the TensorFlow framework and applied to the WIDER dataset [29] on the Nvidia GTX 1050Ti GPU. While the VGG network was trained on various sources from the internet. Experimental results show that the system achieves an accuracy of 94.66%. Cidota and Datcu [30] tried to resolve the early school dropout in Romania by addressing the problem of unequal access to resources, the rural-urban gap, and integration issues. For that, they proposed an augmented reality system for isolated student learning and those with learning difficulties by enabling remote learning and collaborative sessions. The system uses the Microsoft OST-HMD Hololens 2 to show the virtual content provided by the teacher and recognize mathematical expressions using the OCR module.

The development of a comprehensive system that not only records attendance but also integrates features such as parental notifications, secure storage of registration documents, academic performance analysis, and behavior recognition would be an ideal solution for preventing school dropouts. This system would offer a holistic approach, combining real-time data on student engagement, academic trends, and behavioral patterns. By proactively notifying parents and educators of potential concerns, it would enable timely interventions tailored to each student's needs, ultimately fostering a more effective and supportive learning environment while reducing dropout rates.

12.3 DEVELOPING AN IOT-EMBEDDED SYSTEM FOR PREVENTING STUDENT DROPOUT

This work aims to implement a cutting-edge system that leverages digital twin technology to prevent student dropout and ensure safety and privacy. This system will incorporate facial recognition to create a highly robust and secure student attendance solution. Unlike RFID (Radio Frequency Identification), NFC (Near Field Communication), or BLE (Bluetooth Low Energy) tags, which can be easily manipulated by presenting the tags of absent students, a vision-based approach – particularly when leveraging 3D data – offers a much higher level of security and accuracy, making it far more difficult to deceive or bypass. Moreover, an advanced activity recognition framework to detect and flag suspicious behaviors such as bullying or assault is proposed. It will also regulate access to restricted areas, like chemical laboratories, by denying entry to unsupervised students. Furthermore, it ensures the preservation of child privacy and offers real-time alerts to both the school principal and class monitors, enabling swift interventions when necessary.

12.3.1 Overview of Architecture

The general architecture of our suggested framework comprises four major systems: (1) vision-based deep learning attendance system, (2) access control, (3) bullying and assault detection based on activity recognition, and (4) attendance marking and IoT system for data logging and alert notification. The whole system is implemented on the Nvidia Jetson Nano development board. We leverage the GPU CUDA cores of the Tegra X1 SoC integrated into this single-board computer to significantly accelerate computational tasks. Additionally, a 4G SIM communication module, connected through the board's 40-pin GPIO interface, enables the system to send real-time notifications via SMS or email, ensuring prompt communication. Figure 12.1 shows the embedded IoT system for school dropout prevention.

12.3.2 Face Detection and Recognition

When developing a vision-based attendance marking system, ease of use is a top priority, especially in developing countries where access to skilled developers and

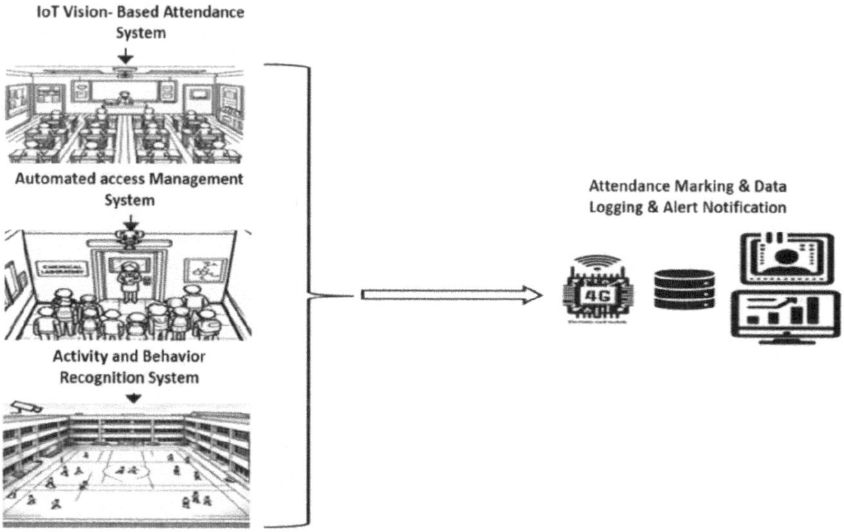

FIGURE 12.1 Architecture of the embedded IoT system for school dropout prevention

specialized hardware may be limited. The challenge intensifies when students need to be added or removed from the system, as this requires updates that should be simple and efficient. Therefore, the system must allow easy modification of information without needing a professional data scientist or advanced hardware for training. At the same time, it should maintain high accuracy with minimal misclassification. Ideally, the solution should be built using an open-source, free library to ensure accessibility and cost-effectiveness.

The Dlib library [31] offers a perfect solution for such a task. As a free, open-source, cross-platform C++ library with a Python wrapper, it strikes an excellent balance between precision and performance [20, 32–34]. Benchmark results demonstrate a recognition accuracy of 99.38% on the widely used Labeled Faces in the Wild dataset [35]. Despite its outstanding accuracy, the execution time of its implemented methods – such as HOG and CNN for face detection, and ResNet [36] CNN for face recognition tends to be longer compared to other techniques and struggles to meet the real-time requirements, especially when targeting embedded systems where resources and power are very constrained [20, 37]. To address this challenge, our embedded system must incorporate a hardware accelerator to significantly boost the performance of computationally intensive CNNs. There are many embedded accelerators: embedded graphics processing units (GPUs), tensor processing units TPUs, digital signal processors DSPs, visual processing units VPUs, application-specific integrated circuits ASICs, field programmable gate arrays FPGAs, neural processing units NPUs, accelerated processing units APUs, and AI SoCs. In this work, an Nvidia-embedded GPU is utilized to accelerate computational tasks. This embedded GPU, featuring 128 Maxwell architecture cores, is fitted into the Tegra X1 SoC module. The SoC is integrated into the Jetson Nano system on module SOM, which is, in turn, mounted on a carrier board (Figure 12.2). The primary reason for selecting this board is its capability to execute and accelerate Dlib and OpenCV libraries on the GPU, significantly improving performance. Additionally, its user-friendly design and comprehensive documentation make it an ideal choice for efficient development and deployment. To take advantage of the GPU, Dlib must be built from source.

The whole attendance system executes in several steps:

- Data Collection

First, photos of the enrolled student are taken with the full name, parent's email, and phone number. The photos should be clear, preferably taken from different angles, and with high resolution. Each student has an appropriate folder with all photos and information on it.

- Image Preprocessing

Image preprocessing is an important step, especially for inference running when conditions like illumination change and image-capturing device defects are uncontrolled. Techniques such as histogram equalization and noise removal enhance feature extraction, leading to improved overall performance.

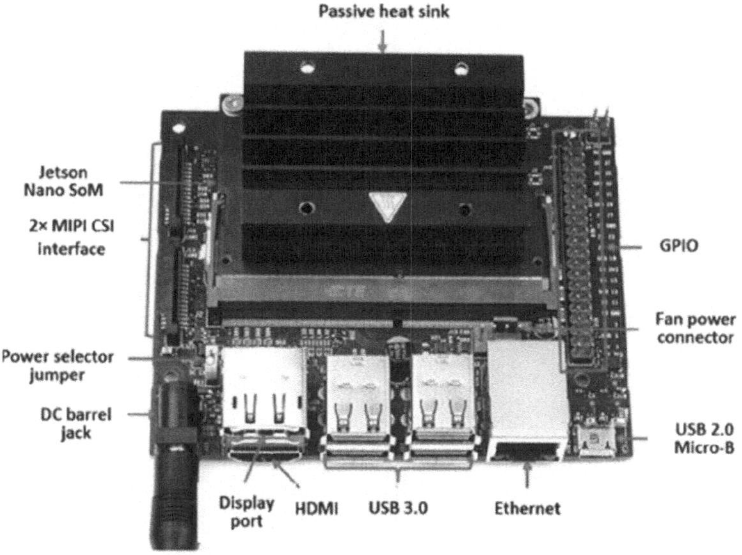

FIGURE 12.2 The Jetson Nano developer kit

- Face Detection

Face detection is a crucial step in face recognition. Faces must be localized in the image to limit the area where the recognition will be applied, thus minimizing false recognition. Two face detectors are implemented on the Dlib: a histogram of oriented gradients (HOG) [21] and a ResNet-SSD [27].

The HOG is a local feature descriptor traditional machine-learning method for object detection. HOG computes histograms of gradient directions in localized regions of an image, capturing objects' shape and appearance. The HOG associated with the SVM classifier is widely used for such tasks and has demonstrated strong accuracy, particularly with smaller datasets. Additionally, its computation time is significantly faster compared to deep learning approaches, making it an efficient choice for many applications.

ResNet-SSD is a deep learning convolutional neural network composed of ResNet [36] architecture and single shot mutltibox detector SSD [27]. The ResNet is used as a backbone for feature extraction. It generated rich and deep feature maps efficiently by introducing residual blocks (skip layers), preventing issues like vanishing gradients in very deep neural networks. Combined with SSD, which predicts bounding boxes and class labels of the objects at various scales and feature map resolutions in a single shot (single forward pass), this method achieves higher accuracy compared to the HOG method, though with increased latency. The developer has the ability to choose between these two implemented methods on the Dlib.

- Face Alignment

Face alignment is applied to help minimize variations caused by different poses, angles, scales, and sizes. It reduces the differences within images of the same person due to varying head poses and expressions. Thereby making the recognition process more invariant to pose, which reduces misclassification. The Dlib library incorporates a face alignment method referred to as the "Dlib facial landmark detector." This method is based on a regression tree ensemble that detects and predicts the locations of 68 facial landmarks, including points around the eyes, eyebrows, nose, mouth, and jawline. It was trained on the 68-point iBUG 300-W facial landmark dataset [38].

- Face Recognition

Face recognition on the Dlib is an embedding-based approach to which face recognition is not applied directly. Instead, it maps an image of a face to a point in a high-dimensional embedding space (typically a 128-dimensional or 512-dimensional vector). This vector, often referred to as an embedding, represents the face in such a way that the distance between two embeddings indicates how similar the faces are. Dlib model works similarly to FaceNet [22] and OpenFace [39]. However, the model is based on a modified ResNet-34 architecture [22], where the classification layer and the softmax layer are removed. In contrast, the global average pooling layer (the new output layer) is modified to generate the 128 embeddings, resulting in a model with 29 Conv layers. A triplot loss is used for training this model; this function ensures that the anchor (an image of an individual or a person) is closer to a positive image (another image of the same individual) than to a negative image (an image of a different person) in the embedding space. Basically, the weights of our model are adjusted in a manner where the distance between the anchor and the positive is minimized, and the distance between the anchor and the negative is maximized by a margin. The 128 embeddings of all enrolled students' faces are stored as a serialized data type using the pickle module.

- Face Identification

The final stage involves face identification and verification. At this point, faces are detected in each new image, and for every detected face, a 128-dimensional embedding (vector) is generated. These embeddings encapsulate the unique facial features necessary for comparison. To identify students, each vector is matched against the previously stored serialized embeddings of the enrolled students. The comparison is performed using the Euclidean distance as a metric, with a predefined tolerance (threshold), to ensure that even the "best match" is within an acceptable similarity range. To further refine the results, the minimum distance to the new face is used to identify the best match. Figure 12.3 illustrates the different stages of face recognition inference in the attendance system.

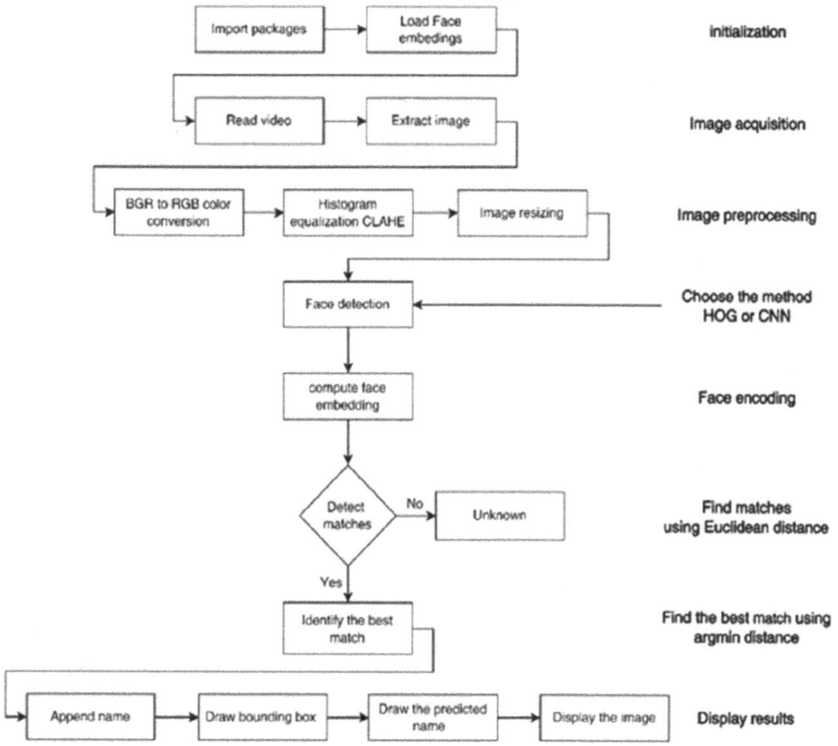

FIGURE 12.3 Illustration of the various stages of face detection and recognition in the attendance system

12.3.3 Automated Access Management

The developed access control system (Figure 12.4) relies on identical face detection and recognition systems discussed in Section 12.3.2. However, the system contains face embeddings of authorized teachers and school staff (Lab engineers, technicians, administrators). The system is mounted in each door laboratory, and access is granted whenever a match is found with the database. The door is actioned via an electromagnetic relay connected to the Jetson Nano GPIO pin and controlled via the Jetson GPIO Library* provided by Nvidia.

12.3.4 Activity Recognitions

As highlighted in Section 12.2.1, numerous studies confirm that bullying and adverse psychological conditions are among the primary drivers of student dropout rates. Persistent harassment, social isolation, and an overall hostile school

* https://github.com/NVIDIA/jetson-gpio.

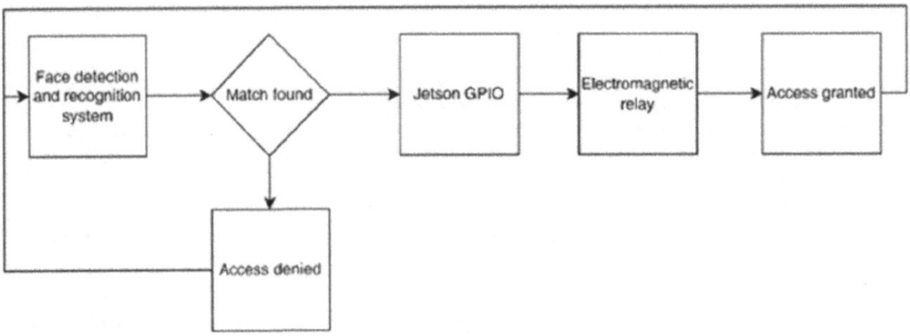

FIGURE 12.4 Flowchart of access control system

environment contribute significantly to students' emotional distress, often leading them to disengage from their education. The inability to cope with such negative experiences can result in a downward spiral, where academic performance declines, self-esteem suffers, and eventually, the decision to leave school becomes inevitable. Addressing these issues is crucial to fostering a supportive educational environment that encourages retention and student well-being. Therefore, developing a system that recognizes activities and alerts the school principal and hall monitors in case of bullying or fighting is crucial. Many techniques involving the use of spatiotemporal information, given the nature of the actions and activities, have been developed in the last two decades. Starting from traditional machine learning methods that use handcrafted spatiotemporal techniques such as (1) spatiotemporal interest point detectors STIP [40, 41], Harris 3D [40], cuboid detector [42], Hessian detector [43], Spatial-Temporal Implicit Shape Model STISM [44], V-FAST [45]. These detectors are often combined with descriptors such as (2) histogram of oriented gradients HOG [21] + histogram of optical flow HOF [46], HOG3D [47], cuboid descriptor [42], extended speeded up robust features ESURF [43], Motion Boundary Histograms MBH [48, 49], 3D-SIFT [50], and local trinary patterns LTP [51]. The combination of detectors and descriptors is followed by a vocabulary builder such as (3) bag of visual words [52–54], which converts image descriptors into a codebook to represent activities. A classifier is applied to learn and predict action/activity classes; commonly used classifiers for action/activity recognition include k-nearest Neighbors (k-NN), SVM, DT and Random Forests, Hidden Markov Models (HMMs) and Gaussian Mixture Models (GMMs). Other traditional methods rely on pose estimation to recognize actions, such as the pictorial structures model [55] and Poselets [56, 57].

With the rise of deep learning methods, new methods have emerged. These methods even automatically extract the relevant features or combine handcrafted features with DNN. Examples of these architectures include spatiotemporal convolutional neural networks (3D-CNNs) [58–60], multi-stream CNNs [61–63], Long-Short-Term Memory (LSTM-CNN) [64–66], Long-Short-Term Memory recurrent neural networks (LSTM-RNN) [67, 68], Temporal convolutional networks TCN

[69, 70], Spatial-Temporal Graph Convolutional Networks (ST-GCN) [71, 72], and Spatiotemporal vision transformers [73–75].

Despite the promising performance of deep learning methods, their application requires access to huge, annotated datasets for effective training. To the best of our knowledge, there are currently no publicly available video or image datasets specifically focused on student bullying or fighting. Furthermore, implementing such systems, even with traditional machine learning methods, on embedded platforms presents significant challenges due to the computational complexity involved. Nevertheless, efforts have been initiated to collect and label relevant datasets, which will facilitate future development in this area.

12.3.5 IoT System (Attendance Marking, Parent's Notification, Immediate Alert, and Data Logging)

The automatic attendance marking system offers several advantages over traditional methods. In conventional systems, issues such as students signing in for absent peers, illegible signatures, incorrect sign-in locations, incomplete or outdated student lists, or missed registration opportunities are common. Additionally, significant time is wasted during each session for roll call. An automated system eliminates these problems, ensuring accuracy, efficiency, and real-time updates. Our attendance marking system, which is associated with the face detection recognition system, will automatically register all the present students in a CSV (Comma-separated values) file each hour. Further, this file will be sent to the school administration via email, and the parents of each absent student will receive an email and SMS notification containing the date and hour when their child is absent.

Communication is ensured via the SIM7600G-H 4G module, as shown in Figure 12.5. This module utilizes the Jetson GPIO library to interface with the board via the USART protocol while employing USB for email transmission over a 4G connection.

To reduce misses (present marked as absent) or false positives (absent marked as present), we adopted the same strategy employed in [23], where the student is marked present if it is recognized more than five times out of ten trials.

12.4 EXPERIMENTAL RESULTS

To achieve a consistent result, the system was mounted in a real environment: a classroom in the Higher Teacher School of Laghouat. The system was placed in the upper front above the whiteboard. A Logitech C920HD was used for image acquisition and connected to the Jetson developer board via USB3. The Logitech has a field of view of 78° and an auto light correction, making it suitable for this task.

Ten students were involved; for each one, ten photos were taken using an iPhone 11 in HD resolution. The first step is the generation of face encodings (embeddings) for all enrolled students, which will be saved as a pickle file to be used later for comparison in the inference.

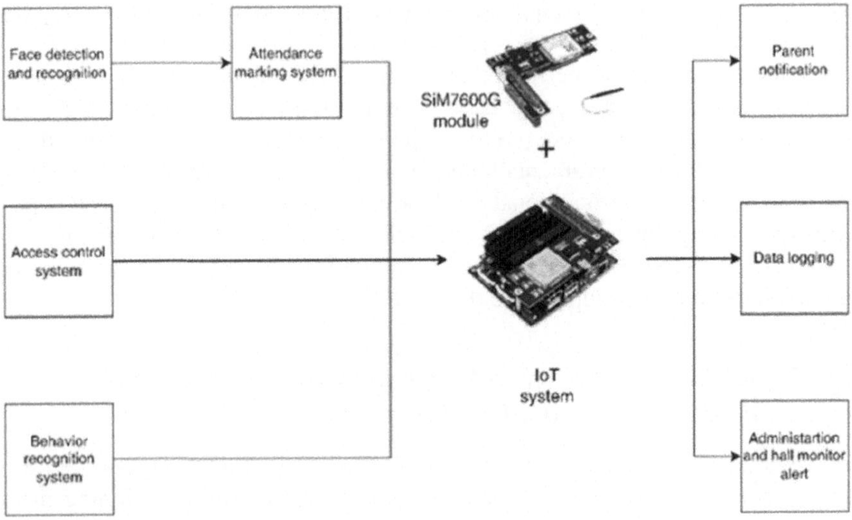

FIGURE 12.5 Flowchart of the Smart IoT vision-based attendance monitoring and activity recognition system for student dropout prevention

To facilitate the addition or removal of students, a graphical user interface is developed using "tkinter" (Tk GUI toolkit) to update the pickle file for non-professionals. Figure 12.6 shows the GUI interface.

For preprocessing, the threshold for contrast limiting in CLAHE was set to 2, while the number of equally sized rectangular tiles was set to 8. Both the HOG and CNN for face detection have been tested.

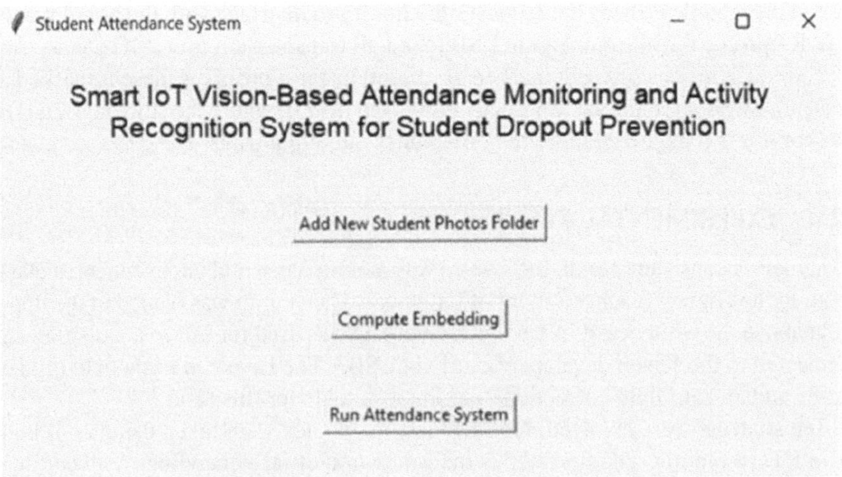

FIGURE 12.6 GUI of the smart IoT vision-based attendance monitoring and activity recognition system for student dropout prevention

For training, results show that CNN face detection took longer than the HOG method on an Intel I7 7700HQ CPU, while the results were similar on the Jetson Nano GPU (Dlib built with CUDA support) (Table 12.1). For inference running, the HOG method is faster than the CNN on the CPU because the HOG has a low computational complexity compared to the CNN; additionally, the CPU has limited parallelism capabilities to accelerate convolutional neural networks. On the other hand, running the inference on the Jetson Nano GPU shows that the CNN execution is faster than the HOG method due to CUDA and Cudnn libraries that have optimized implementations of CNN layers for parallel execution. Therefore, the CNN method was employed for face detection.

For the matching test, an empirical threshold of 0.47 was employed. Using this value, the complete system achieves an average accuracy for face recognition of 92.12% and an F1 score of 93.98%. Figure 12.7 shows the result of the attendance system based on face detection and recognition. For the attendance marking system, as mentioned earlier, the student is marked present if it is recognized more than five times out of ten trials, which increases the accuracy of attendance marking further.

12.5 CONCLUSION AND FUTURE WORK

In this chapter, we have presented for digital twins the design and implementation of a full dropout prevention system composed of four integrated subsystems: (1) a vision-based attendance detection system, (2) an access control system, (3) a behavior and activity recognition system, and (4) an IoT attendance marking and notification system. The IoT vision-based attendance detection and access control systems

TABLE 12.1
Execution Time of Face Detection and Recognition (Training/Inference)

Jetson Nano (GPU)	Laptop (CPU) I7 7700HQ	The Model
41.91 seconds	12.87 seconds	Training time (HOG)
42.27 seconds	229.81 seconds	Training time (CNN)
3.05 seconds	0.50 seconds	Inference time (HOG)
1.58 seconds	0.84 seconds	Inference time (CNN)

FIGURE 12.7 Results of face detection and recognition system

were successfully deployed on the Jetson Nano embedded platform, leveraging its GPU capabilities to achieve high accuracy and computational efficiency. Despite this success, the behavior recognition system remains in development, hindered by the lack of an available public dataset. Addressing this gap will be the focus of our future work. Moreover, the Jetson Nano's limited processing power restricts its ability to handle additional deep neural networks, especially activity recognition networks, prompting us to explore an upgrade to the Jetson Orin family, which offers enhanced performance. Further research and experimentation will be necessary to validate the scalability, robustness, and overall efficiency of the proposed system in real-world scenarios.

Availability of Data

The datasets generated during and/or analyzed during the current study are available from the corresponding author upon reasonable request.

REFERENCES

1. UNESCO, "250 Million Children out-of-school: What You Need to Know about UNESCO's Latest Education Data," *Unesco.org*, Sep. 19, 2023. Available at: https://www.unesco.org/en/articles/250-million-children-out-school-what-you-need-know-about-unescos-latest-education-data
2. A. Reschly and K. S. Patton, *School Dropout*, Springer eBooks, pp. 5692–5695, Jan. 2014. https://doi.org/10.1007/978-94-007-0753-5_2598
3. J. Ogresta, I. Rezo, P. Kožljan, M.-H. Paré, and M. Ajduković, "Why Do We Drop Out? Typology of Dropping Out of High School," *Youth & Society*, 0044118X2091843, 53(6), 2020. https://doi.org/10.1177/0044118x20918435
4. N. Hoff, A. Olson, and R. L. Peterson, "Dropout Screening & Early Warning," *Building & Sustaining*, 2015.
5. I. Archambault, M. Janosz, E. Olivier, and V. Dupéré, "Student Engagement and School Dropout: Theories, Evidence, and Future Directions," in *Handbook of Research on Student Engagement*, 1, pp. 331–355, 2022. https://doi.org/10.1007/978-3-031-07853-8_16
6. H. Hirose, "Attendance to Lectures is Crucial in Order not to Drop Out," in *2018 7th International Congress on Advanced Applied Informatics*, Jul. 2018. https://doi.org/10.1109/iiai-aai.2018.00045
7. R. Samuel and K. Burger, "Negative life events, self-efficacy, and social support: Risk and protective factors for school dropout intentions and dropout," *Journal of Educational Psychology*, 112(5), pp. 973–986, 2019. https://doi.org/10.1037/edu0000406
8. Ş. Şahin, Z. Arseven, and A. Kılıç, "Causes of Student Absenteeism and School Dropouts," *International Journal of Instruction*, 9(1), 195–210, 2016. https://doi.org/10.12973/iji.2016.9115a. Available: http://www.e-iji.net/dosyalar/iji_2016_1_14.pdf
9. S. Lamb and E. Al, *School dropout and completion: international comparative studies in theory and policy.* Dordrecht: Springer, 2011.
10. R. W. Rumberger, "The economics of high school dropouts," *The Economics of Education*, 149–158, 2020. https://doi.org/10.1016/b978-0-12-815391-8.00012-4

11. M. Psyridou, F. Prezja, M. Torppa, M.-K. Lerkkanen, A.-M. Poikkeus, and K. Vasalampi, "Machine learning predicts upper secondary education dropout as early as the end of primary school," *Scientific Reports*, 14(1), 2024. https://doi.org/10.1038/s41598-024-63629-0
12. L. Earving, J. A. Castán-Rocha, S. Ibarra-Martínez, J. D. Terán-Villanueva, M. G. Treviño-Berrones, and J. Laria-Menchaca, "Cluster Analysis Using k-Means in School Dropout," in *Studies in Big Data*, pp. 3–16, Jan. 2023. https://doi.org/10.1007/978-3-031-38325-0_1
13. I. H. Witten, E. Frank, M. A. Hall, and C. J. Pal, *Data mining: practical machine learning tools and techniques*. Amsterdam: Morgan Kaufmann, 2017.
14. P. Alejandra Cuevas-Chávez, S. Narciso, E. Sánchez-Jiménez, Itzel Celerino Pérez, Yasmín Hernández, and J. Ortiz-Hernandez, "School Dropout Prediction with Class Balancing and Hyperparameter Configuration," *Lecture Notes in Computer Science*, 12–20, 2024. https://doi.org/10.1007/978-3-031-51940-6_2
15. Lucía Morán, "higher-education-predictors-of-student-retention," Kaggle.com, 2023. Available at: https://www.kaggle.com/datasets/lucamorn/higher-education-predictors-of-student-retention [Accessed on September 19, 2024].
16. F. A. d. S. Freitas et al., "IoT System for School Dropout Prediction Using Machine Learning Techniques Based on Socioeconomic Data," *Electronics*, 9(10), 1613, Oct. 2020. https://doi.org/10.3390/electronics9101613
17. M. Zerkouk, M. Mihoubi, B. Chikhaoui, and S. Wang, "A machine learning based model for student's dropout prediction in online training," *Education and Information Technologies*, 29, 2024. https://doi.org/10.1007/s10639-024-12500-w
18. O. Bulut, T. Wongvorachan, S. He, and S. Lee, "Enhancing high-school dropout identification: a collaborative approach integrating human and machine insights," *Discover Education*, 3(1), 2024. https://doi.org/10.1007/s44217-024-00209-4
19. K. Niu, G. Lu, X. Peng, Y. Zhou, J. Zeng, and K. Zhang, "CNN autoencoders and LSTM-based reduced order model for student dropout prediction," *Neural Computing and Applications*, 35(30), 22341–22357, 2023. https://doi.org/10.1007/s00521-023-08894-2
20. A. Anand, V. Jha, Lavanya Sharma, An Improved Local Binary Patterns Histograms Techniques for Face Recognition for Real Time Application, *in International Journal of Recent Technology and Engineering*, 8,(2S7), pp. 524–529, July 2019.
21. N. Dalal and B. Triggs, "Histograms of Oriented Gradients for Human Detection," in *2005 IEEE Computer Society Conference on Computer Vision and Pattern Recognition (CVPR'05)*, Vol. 1, pp. 886–893, 2005. https://doi.org/10.1109/cvpr.2005.177
22. F. Schroff, D. Kalenichenko, and J. Philbin, "FaceNet: A unified embedding for face recognition and clustering," in *2015 IEEE Conference on Computer Vision and Pattern Recognition (CVPR)*, 2015. https://doi.org/10.1109/cvpr.2015.7298682
23. W. Kuang and A. Baul, "A Real-time Attendance System Using Deep-learning Face Recognition," in *2020 ASEE Virtual Annual Conference Content Access Proceedings*, 2020. https://doi.org/10.18260/1-2--33949
24. P. Viola and M. Jones, "Rapid object detection using a boosted cascade of simple features," in *Proceedings of the 2001 IEEE Computer Society Conference on Computer Vision and Pattern Recognition. CVPR 2001*, Vol. 1, 2001. https://doi.org/10.1109/cvpr.2001.990517
25. H. El Mrabet and A. Ait Moussa, "IoT-School Attendance System Using RFID Technology," *International Journal of Interactive Mobile Technologies (iJIM)*, 14(14), 95, 2020. https://doi.org/10.3991/ijim.v14i14.14625

26. J. Joshan Athanesious, Vanitha, S. Adithya, C. A. Bhardwaj, J. S. Lamba, and A. V. Vaidehi, "Deep Learning Based Automated Attendance System," *Procedia Computer Science*, 165, 307–313, 2019. https://doi.org/10.1016/j.procs.2020.01.045
27. W. Liu et al., "SSD: Single Shot MultiBox Detector," in *Computer Vision – ECCV 2016*, Vol. 9905, pp. 21–37, 2016. https://doi.org/10.1007/978-3-319-46448-0_2. Available at: https://arxiv.org/abs/1512.02325
28. K. Simonyan and A. Zisserman, "Very Deep Convolutional Networks for Large-Scale Image Recognition," *Computer Science*, 2014. https://doi.org/10.48550/arXiv.1409.1556
29. S. Yang, P. Luo, Chen Change Loy, and X. Tang, "WIDER FACE: A Face Detection Benchmark," *arXiv* (Cornell University), Nov. 2015. https://doi.org/10.48550/arxiv.1511.06523
30. M. A. Cidota and Dragos Datcu, "Remote Assistance System in Augmented Reality for Early School Dropout Prevention," in *2020 IEEE International Symposium on Mixed and Augmented Reality Adjunct (ISMAR-Adjunct)*, Nov. 2020. https://doi.org/10.1109/ismar-adjunct51615.2020.00091
31. D. E. King, "Dlib-ml: A Machine Learning Toolkit," *Journal of Machine Learning Research*, 10(60), 1755–1758, 2024, Available at: http://jmlr.org/papers/v10/king09a.html [Accessed on September 19, 2024].
32. M. Xu, D. Chen, and G. Zhou, "Real-Time Face Recognition Based on Dlib," in *Lecture Notes in Electrical Engineering*, pp. 1451–1459, 2020. https://doi.org/10.1007/978-981-15-5959-4_177
33. N. Boyko, O. Basystiuk, and N. Shakhovska, "Performance Evaluation and Comparison of Software for Face Recognition, Based on Dlib and Opencv Library," *2018 IEEE Second International Conference on Data Stream Mining & Processing (DSMP), IEEE Xplore*, Lviv, Ukraine, 2018, pp. 478–482, https://doi.org/10.1109/DSMP.2018.8478556. keywords: {Face;Face recognition;Libraries;Computer vision;Support vector machines;Task analysis;Estimation;computer vision;face recognition;algorithms;performance;dlib;openCV;machine learning;HOG;face landmark estimation;DCNN;SVM}
34. B. Handaga, B. Murtiyasa, and J. Wantoro, "Attendance System based on Deep Learning Face Recognition without Queue," in *2019 Fourth International Conference on Informatics and Computing (ICIC)*, Oct. 2019. https://doi.org/10.1109/icic47613.2019.8985697
35. G. B. Huang, M. Mattar, T. Berg, and E. Learned-Miller, "Labeled Faces in the Wild: A Database for Studying Face Recognition in Unconstrained Environments," in *Workshop on Faces in "Real-Life" Images: Detection, Alignment, and Recognition*, Oct. 2008.
36. K. He, X. Zhang, S. Ren, and J. Sun, "Deep Residual Learning for Image Recognition," in *Proceedings of the IEEE Conference on Computer Vision And Pattern Recognition*, Dec. 2015. https://doi.org/10.48550/arxiv.1512.03385
37. K. Sehairi and T. Bouwmans, "Benchmarking Different Deep Learning Optimizer and Runtime APIs for Face Detection and Recognition Applications on Different Embedded Systems," Manuscript submitted for publication].
38. C. Sagonas, E. Antonakos, G. Tzimiropoulos, S. Zafeiriou, and M. Pantic, "300 Faces In-The-Wild Challenge: database and results," *Image and Vision Computing*, 47, 3–18, 2016. https://doi.org/10.1016/j.imavis.2016.01.002.
39. B. Amos, B. Ludwiczuk, and M. Satyanarayanan, "OpenFace: A general-purpose face recognition library with mobile applications," 2016. Available at: https://elijah.cs.cmu.edu/DOCS/CMU-CS-16-118.pdf
40. I. Laptev, "On Space-Time Interest Points," *International Journal of Computer Vision*, 64(2–3), 107–123, 2005. https://doi.org/10.1007/s11263-005-1838-7

41. D. Das Dawn and S. H. Shaikh, “A comprehensive survey of human action recognition with spatio-temporal interest point (STIP) detector,” *The Visual Computer*, 32(3), 289–306, 2015. https://doi.org/10.1007/s00371-015-1066-2
42. P. Dollar, V. Rabaud, G. Cottrell, and S. Belongie, “Behavior Recognition via Sparse Spatio-Temporal Features,” in *2005 IEEE International Workshop on Visual Surveillance and Performance Evaluation of Tracking and Surveillance*, 2019. https://doi.org/10.1109/vspets.2005.1570899.
43. G. Willems, T. Tuytelaars, and L. Van Gool, “An Efficient Dense and Scale-Invariant Spatio-Temporal Interest Point Detector,” in *Lecture Notes in Computer Science*, pp. 650–663, 2008. https://doi.org/10.1007/978-3-540-88688-4_48.
44. G. Yu, J. Yuan, and Z. Liu, “Predicting human activities using spatio-temporal structure of interest points,” in *Proceedings of the 20th ACM international conference on Multimedia*, Oct. 2012. https://doi.org/10.1145/2393347.2396380
45. T.-H. Yu, Tae Yong Kim, and R. Cipolla, “Real-time Action Recognition by Spatiotemporal Semantic and Structural Forests,” in *BMVC*, Vol. 2, no. 5, Sep. 2010. https://doi.org/10.5244/c.24.52
46. I. Laptev, M. Marszalek, C. Schmid, and B. Rozenfeld, “Learning realistic human actions from movies,” in *2008 IEEE Conference on Computer Vision and Pattern Recognition*, Jun. 2008. https://doi.org/10.1109/cvpr.2008.4587756.
47. A. Klaser, M. Marszałek, and C. Schmid, “A Spatio-Temporal Descriptor Based on 3D-Gradients,” in *BMVC 2008 - 19th British Machine Vision Conference*, Sep. 2008, p. 275-1.
48. N. Dalal, B. Triggs, and C. Schmid, “Human Detection Using Oriented Histograms of Flow and Appearance,” in *Computer Vision – ECCV 2006*, pp. 428–441, 2006. https://doi.org/10.1007/11744047_33
49. H. Wang, A. Klaser, C. Schmid, and C.-L. Liu, “Action recognition by dense trajectories,” in *CVPR 2011*, Jun. 2011. https://doi.org/10.1109/cvpr.2011.5995407
50. P. Scovanner, S. Ali, and M. Shah, “A 3-dimensional sift descriptor and its application to action recognition,” in *Proceedings of the 15th international conference on Multimedia - MULTIMEDIA ’07*, 2007. https://doi.org/10.1145/1291233.1291311
51. Lahav Yeffet and L. Wolf, “Local Trinary Patterns for human action recognition,” in *International Conference on Computer Vision*, Sep. 2009. https://doi.org/10.1109/iccv.2009.5459201
52. G. Csurka, C. R. Dance, L. Fan, J. Willamowski, and C. Bray, “Visual categorization with bags of keypoints,” in *European Conference on Computer Vision*, Vol. 1, p. 22, Jan. 2004.
53. A. Kovashka and K. Grauman, “Learning a hierarchy of discriminative space-time neighborhood features for human action recognition,” in *2010 IEEE computer society conference on computer vision and pattern recognition*, Jun. 2010, pp. 2046–2053. https://doi.org/10.1109/CVPR.2010.5539881
54. X. Peng, L. Wang, X. Wang, and Y. Qiao, “Bag of visual words and fusion methods for action recognition: Comprehensive study and good practice,” *Computer Vision and Image Understanding*, 150, 109–125, 2016. https://doi.org/10.1016/j.cviu.2016.03.013
55. P. F. Felzenszwalb and D. P. Huttenlocher, “Pictorial Structures for Object Recognition,” *International Journal of Computer Vision*, 61(1), 55–79, Jan. 2005. https://doi.org/10.1023/b:visi.0000042934.15159.49
56. M. Raptis and L. Sigal, “Poselet Key-Framing: A Model for Human Activity Recognition,” in *Proceedings of the IEEE Conference on Computer Vision and Pattern Recognition*, Jun. 2013. https://doi.org/10.1109/cvpr.2013.342

57. L. Tao and R. Vidal, "Moving Poselets: A Discriminative and Interpretable Skeletal Motion Representation for Action Recognition," in *Proceedings of the IEEE international conference on computer vision workshops*, Dec. 2015. https://doi.org/10.1109/iccvw.2015.48
58. S. Ji, W. Xu, M. Yang, and K. Yu, "3D Convolutional Neural Networks for Human Action Recognition," *IEEE Transactions on Pattern Analysis and Machine Intelligence*, 35(1), 221–231, 2013. https://doi.org/10.1109/tpami.2012.59
59. D. Tran, L. Bourdev, R. Fergus, L. Torresani, and M. Paluri, "Learning Spatiotemporal Features with 3D Convolutional Networks," in *Proceedings of the IEEE international conference on computer vision*, Dec. 2014. https://doi.org/10.48550/arxiv.1412.0767
60. K. Hara, H. Kataoka, and Y. Satoh, "Can Spatiotemporal 3D CNNs Retrace the History of 2D CNNs and ImageNet?," *IEEE Xplore*, Jun. 01, 2018. https://doi.org/10.1109/CVPR.2018.00685
61. C. Feichtenhofer, A. Pinz, and A. Zisserman, "Convolutional Two-Stream Network Fusion for Video Action Recognition," in *Proceedings of the IEEE conference on computer vision and pattern recognition*, Sep. 2016. https://doi.org/10.48550/arXiv.1604.06573
62. Z. Tu et al., "Multi-stream CNN: Learning representations based on human-related regions for action recognition," *Pattern Recognition*, 79, 32–43, 2018. https://doi.org/10.1016/j.patcog.2018.01.020
63. H. Wang, B. Yu, J. Li, L. Zhang, and D. Chen, "Multi-Stream Interaction Networks for Human Action Recognition," *IEEE Transactions on Circuits and Systems for Video Technology*, 32(5), 3050–3060, 2022. https://doi.org/10.1109/tcsvt.2021.3098839
64. S. Hochreiter and J. Schmidhuber, "Long Short-Term Memory," *Neural Computation*, 9(8), 1735–1780, 1997. https://doi.org/10.1162/neco.1997.9.8.1735
65. K. Xia, J. Huang, and H. Wang, "LSTM-CNN Architecture for Human Activity Recognition," *IEEE Access*, 8, 56855–56866, 2020. https://doi.org/10.1109/access.2020.2982225
66. L. Wang, Y. Xu, J. Cheng, H. Xia, J. Yin, and J. Wu, "Human Action Recognition by Learning Spatio-Temporal Features With Deep Neural Networks," *IEEE Access*, 6, 17913–17922, 2018. https://doi.org/10.1109/access.2018.2817253.
67. A. Murad and J.-Y. Pyun, "Deep Recurrent Neural Networks for Human Activity Recognition," *Sensors*, 17, 11, 2556, 2017. https://doi.org/10.3390/s17112556
68. D. Singh et al., "Human Activity Recognition Using Recurrent Neural Networks," in *Lecture Notes in Computer Science*, pp. 267–274, 2017. https://doi.org/10.1007/978-3-319-66808-6_18
69. C. Lea, M. D. Flynn, R. Vidal, A. Reiter, and G. D. Hager, "Temporal Convolutional Networks for Action Segmentation and Detection," in *2017 IEEE Conference on Computer Vision and Pattern Recognition (CVPR)*, Jul. 2017. https://doi.org/10.1109/cvpr.2017.113
70. T. Y. Kim and A. Reiter, "Interpretable 3D Human Action Analysis with Temporal Convolutional Networks," *2017 IEEE Conference on Computer Vision and Pattern Recognition* Workshops (CVPRW), Honolulu, HI, USA, 2017, pp. 1623–1631. https://doi.org/10.1109/CVPRW.2017.207. keywords: {Three-dimensional displays;Solid modeling;Skeleton; Feature extraction; Computational modeling;Activity recognition}.
71. S. Yan, Y. Xiong, and D. Lin, "Spatial Temporal Graph Convolutional Networks for Skeleton-Based Action Recognition," *arXiv* (Cornell University), Jan. 2018. https://doi.org/10.48550/arxiv.1801.07455

72. Z. Chen, S. Li, B. Yang, Q. Li, and H. Liu, "Multi-Scale Spatial Temporal Graph Convolutional Network for Skeleton-Based Action Recognition," *Proceedings of the AAAI Conference on Artificial Intelligence*, 35(2), 1113–1122, 2021. https://doi.org/10.1609/aaai.v35i2.16197
73. A. Ulhaq, N. Akhtar, G. Pogrebna, and A. Mian, "Vision Transformers for Action Recognition: A Survey," *arXiv.org*, Sep. 2022. https://doi.org/10.48550/arXiv.2209.05700
74. Naga et al., "SPARTAN: Self-supervised Spatiotemporal Transformers Approach to Group Activity Recognition," *arXiv* (Cornell University), Jun. 2023. https://doi.org/10.1109/cvprw59228.2023.00544
75. H. Han, H. Zeng, L. Kuang, X. Han, and H. Xue, "A human activity recognition method based on Vision Transformer," *Scientific Reports*, 14(1), 15310, 2024. https://doi.org/10.1038/s41598-024-65850-3

13 Emerging Application of Digital Twin Technology toward Development of Smart Infrastructure

Jimmy Mehta, Yaman Hooda, and Yati Aggarwal

13.1 INTRODUCTION

A Digital Twin is a virtual representation of a physical asset, such as a bridge or building. While the technology behind Digital Twins is relatively new, the concept of digitally replicating objects dates back to the 1960s. NASA, for instance, used digital replicas to simulate space conditions during the Apollo 13 mission. Digital Twins can be created using various methods, including 3D design models or reality capture scans, to which sensors can be attached. The creation process varies depending on the replicated asset – a bridge, pipeline, or wastewater treatment plant (WWTP). For example, drones can scan and replicate bridges, LiDAR lasers can map pipelines, and sensors placed throughout a WWTP can simulate operations [1]. Digital Twins have significantly transformed civil engineering by enhancing inspections, routine maintenance, and real-time monitoring. Engineers commonly use two types of Digital Twins: (1) a virtual, static 3D model that captures a snapshot at a specific time, and (2) a virtual, fully integrated model that monitors and controls a facility, such as a WWTP, remotely. These digital replicas allow engineers to manage infrastructure assets – such as roads, bridges, pipelines, and WWTPs – ensuring timely inspections and maintenance to keep them in optimal condition [2].

Computational technology advancements, especially in digitalization and data management, have propelled Industry 4.0, a pivotal function for which Digital Twins (DTs) are essential. With two-thirds of big industrial organizations predicted to deploy at least one DT during the next ten years, the worldwide DT market, estimated at $3.8 billion in 2019, is forecast to grow to $35.8 billion by 2025 and enhance operational efficiency by 10%. A digital twin is, to put it simply, an ever-updated virtual copy of a real system or item. The purpose of this technical chapter is to support the current endeavors in civil engineering to define the notion of digital twins [3]. The analysis stresses the importance of maintaining a clear distinction between the real and virtual worlds by focusing on objective, physical descriptors unrelated to

 DOI: 10.1201/9781003582489-13

any specific modeling framework. Additionally, it highlights that updating a model involves two distinct activities: parameter updating (Type I) and formulation updating (Type II). The latter, which deals with adjusting the model's structure, is identified as an area requiring further research. DTs in the Architecture, Engineering, and Construction (AEC) sector remain mostly in the prototype phase, lacking standard protocols and guidelines for their widespread implementation [4]. This challenge is more pronounced for legacy infrastructure, which often lacks a digital counterpart, unlike newly constructed assets. The UK's National Digital Twin Programme defines DTs as realistic digital representations of assets, processes, or systems. Its crucial feature is the synchronization of real-world data with its digital counterpart. This synchronization can be active, such as in cybernetic systems directly controlling physical assets, or passive, informing decision-making processes like building maintenance. DTs can represent anything from a single component to complex ecosystems, aiming to connect urban systems to create a National Digital Twin [5].

Developing common standards and tailored processes is essential to establishing DTs in AEC. This chapter proposes a generic workflow for creating DTs for existing assets, demonstrated through a case study of the Clifton Suspension Bridge in the UK [6]. In the implementation of Digital Twins (DTs) for civil infrastructure, the Finite Element Method (FEM) is one of the most widely utilized numerical techniques. FEM-based commercial software like ABAQUS is commonly used to develop the virtual twin, which represents the physical asset. Parameters for the virtual twin are derived from the physical structure, and after validation, the FEM model generates substantial datasets that interact with real-world measurement data. In addition to modeling mechanical properties, FEM can also establish electromechanical coupled models that output both mechanical characteristics, such as displacement and load, and electrical data, such as voltage or charge, from embedded sensors. For example, researchers have developed multi-physics models using FEM to simulate wave propagation in composite structures with piezoelectric transducers, which allows real-time structural damage detection [7]. However, a major limitation of FEM-based virtual twins is the substantial computational power required, making them unsuitable for real-time simulation due to constraints in size and time step [8, 9].

A data-driven approach is another methodology for developing digital twins, where models are created and parameterized based on data collected from the physical asset. These models, often combined with machine learning (ML) and artificial intelligence (AI), provide a higher level of accuracy compared to traditional simulation methods [10]. By integrating ML and AI, simulations can capture complex phenomena that conventional models struggle to replicate, offering deeper insights into system operations. However, human oversight is still needed to fully automate these processes [11]. The use of 3D laser scanning has also become prominent in civil infrastructure modeling. This non-contact measurement technique rapidly collects surface [12] topography data, with systems classified as aerial, mobile, or terrestrial, depending on the sensor's location during acquisition. Terrestrial laser scanning (TLS) systems, known for their sub-millimeter precision, speed, and cost-efficiency, are increasingly used for inspection and digital modeling tasks [13, 14]. The data collected by TLS is integrated with BIM models, enabling comprehensive digital management of infrastructure and detection of deviations from design [15, 16].

Another emerging approach involves using Unmanned Aerial Vehicles (UAVs), which leverage advanced camera sensing technologies to capture high-resolution images of civil infrastructure. UAVs follow predefined flight paths, capturing multiple images from different angles, which are then processed into 3D point clouds for virtual twin creation. This method allows for efficient remote data collection, though challenges remain regarding accuracy standards and blind spots that require manual data collection [17–19]. Infrastructure management involves the supervision and optimization of vital systems and facilities that support daily life, such as transportation networks, utilities, and public services. Efficient management of these complex systems requires constant monitoring, regular maintenance, and ongoing innovation to meet growing demands and mitigate risks. Digital twins in infrastructure management offer significant benefits in several key areas:

1.1. **Real-Time Monitoring**: Digital twins aggregate data from multiple sources to provide a live view of infrastructure assets, allowing stakeholders to instantly track the status and performance of physical systems.
1.2. **Predictive Maintenance**: With the help of advanced analytics, digital twins enable managers to predict potential system failures or inefficiencies, optimizing performance and energy consumption.

Figure 13.1 discusses the image highlights key areas where Digital Twins are applied in smart infrastructure and urban planning:

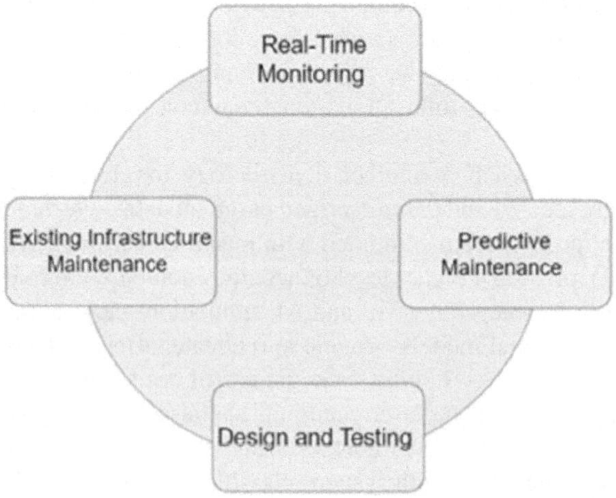

FIGURE 13.1 Key areas of application of Digital Twins in smart infrastructure and urban planning

1. **Real-Time Monitoring**: Enables continuous observation and data collection to optimize infrastructure performance and detect issues.
2. **Predictive Maintenance**: Uses predictive analytics to foresee and prevent equipment failures, minimizing downtime.
3. **Design and Testing**: Facilitates virtual simulations for designing and testing infrastructure before actual implementation.
4. **Existing Infrastructure Maintenance**: Helps maintain and upgrade current infrastructure efficiently by simulating conditions and performance.

1.3. **Design and Testing**: Acting as virtual testbeds, digital twins enable simulations and validation of infrastructure projects, which enhances design efficiency, reduces costs, and shortens development timelines.

1.4. **Existing Infrastructure Maintenance**: Continuous monitoring through digital twins minimizes operational interruptions by ensuring timely maintenance and repairs.

13.2 BASIC CONCEPT: IMPACT OF DIGITAL TWINS ON INFRASTRUCTURE MANAGEMENT

1. **Transportation (Roads, Bridges, and Railways)**: Digital twins provide real-time monitoring, predictive maintenance, and traffic optimization. For instance, a digital twin of a bridge can simulate the effects of different loads, helping engineers identify and address potential structural issues before they become critical.
2. **Renewable Energy Systems**: Digital twins enable detailed simulations of turbines, generators, and entire energy grids, monitoring equipment performance, detecting failures, and optimizing energy production.
3. **Smart Water and Wastewater Management**: By monitoring water quality and predicting equipment failures, digital twins can help optimize operations, such as adjusting chemical dosages and improving sewage flow, thereby enhancing public health and reducing environmental impacts.
4. **Buildings and Facilities**: Digital twins visualize layouts, assess structural integrity, and optimize energy use. After a building becomes operational, these virtual models assist facility managers in adjusting energy consumption, managing occupancy patterns, and scheduling maintenance based on real-time data.

Incorporating technologies like Lidar, radar, satellite data, and InSAR (Interferometric Synthetic Aperture Radar) significantly enhances the capabilities of digital twins by providing detailed earth observation data. Lidar uses laser pulses to create accurate 3D maps of terrain and urban environments, while radar systems detect objects and measure changes under various weather conditions, crucial for monitoring land deformation, water levels, and vegetation. Satellite data, including imagery and environmental metrics, supports global real-time monitoring, climate modeling, and disaster response. InSAR technology tracks minute ground movements, such as subsidence or tectonic activity, by analyzing radar data from multiple satellite passes.

Together, these technologies allow digital twins to simulate and visualize the Earth's dynamics, both natural and human, informing environmental policies and projects like Digital Twin Antarctica, which monitors meltwater movement on the Antarctic ice sheet. IoT sensors are vital for creating digital twins, capturing data such as temperature, pressure, and motion, and providing real-time reflections of physical conditions. Visualization platforms and simulation software, often enhanced by AI, form the digital twin ecosystem's interface, presenting insights through user-friendly platforms like Proqio. This tool provides real-time insights and management solutions for a range of infrastructure applications, from dams to bridges, integrating data throughout the project lifecycle. Proqio's digital twins, complete with 3D views and dynamic heat maps, transform raw data into actionable insights, bridging technology and efficiency in the construction industry as shown in Figure 13.2.

The framework depicts the use of the Finite Element Method (FEM) to create a digital twin for a bridge application. First, vibration tests on a physical bridge generate measured data. This data is used to fine-tune a pre-trained Convolutional Neural Network (CNN), which has been initially trained using numerical bridge models (1 or 2) from the digital twin. The target CNN is then applied for test detection, continuously improving through transfer learning as shown in Figure 13.3.

13.3 PRESENT SCENARIO

13.3.1 Digital Twin and Its Applications in Various Industries

a. *Construction Industry:* Digital twins in construction can provide answers to these questions (Table 13.1). They allow companies to simulate various

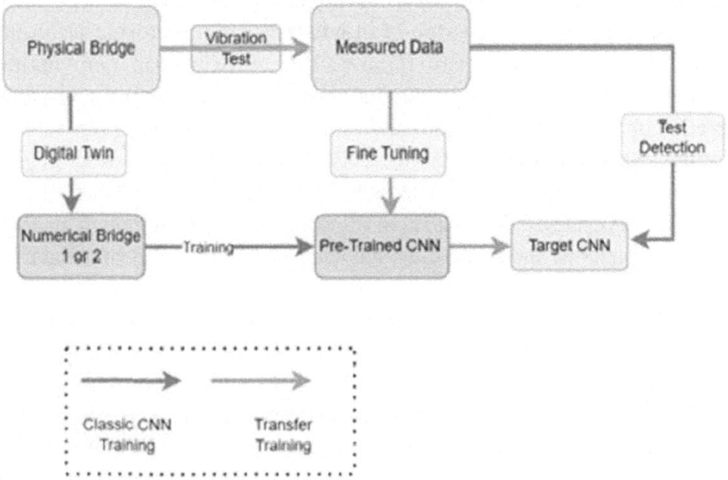

FIGURE 13.2 Structure for creating a digital twin in a bridge application using the Finite Element Method (FEM) technique

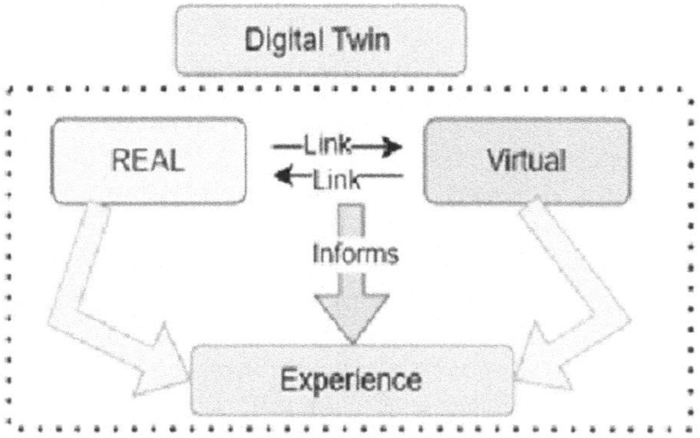

FIGURE 13.3 Typical Digital Twin (DT) components: the real, the virtual, and the link (considering the architecture, engineering, and construction sector)

scenarios – such as design modifications, weather impacts, or even security threats – without interrupting the ongoing construction process. These virtual models offer a real-time, detailed view of how an asset is designed and performed. For example, digital twins enable construction companies to visualize lighting within each room or hall, assess space utilization, and much more. Digital twins promote more sustainable construction practices by helping companies optimize designs, materials, and processes. They play a crucial role in reducing waste and enhancing efficiency.

Digital Twin Example: Crossrail, London

Crossrail, a major infrastructure project in London, has harnessed the power of digital twins to visualize the final product. The technology allowed engineers to monitor progress in real time and identify potential issues early. By simulating and optimizing the construction process, they streamlined workflows and minimized resource waste, ensuring a smoother, more efficient build.

b. *Manufacturing Industry*: The use of digital twins in manufacturing is on the rise, driven by their ability to boost operational efficiency. By creating a virtual model of an industrial process, manufacturers can quickly analyze data, track performance trends over time, and enhance their workflows. One practical application of digital twins is in production line monitoring. Companies leverage these models to track key performance indicators such as temperature, pressure, speed, and environmental conditions. Digital twins can alert manufacturers to potential issues or defects by identifying any irregularities in production. Additionally, digital twins play a critical role in predictive maintenance. Identifying early signs of equipment problems helps prevent costly breakdowns, minimize downtime, and improve product reliability.

TABLE 13.1
Digital Twin Technology Applied in Different Research

S. No	Section Application	Section Application
1	Smart cities and urban spaces	• Quantifying and detecting concrete fractures using transformer-based neural networks and UAVs [20] • Using UAV visual surveys to forecast building performance [21] • Convolutional neural networks-based enhanced identification of structural deterioration [22] • Post-quake structure condition evaluation using digital twin and unmanned aerial vehicle [23] • Using finite element models and deep neural networks to locate and measure truss deterioration [24]
2	Transportation	• Life cycle assessment (LCA) for Precast Advanced Track (PCAT) slab systems is evaluated using digital twins [25] • Using DT technology for resilience optimization and railway maintenance [26] • A switch machine maintenance concept based on digital twins [27] • An autonomous feedback-based schedule optimization solution for railway delays [28]
3	Energy system	• A DT-based methodology to identify the most effective energy-saving tactics and technologies for buildings that are already in use [29] • Managing energy resources to optimize use, analyze it, and meet sustainable development objectives [30] • A linked simulation that combines the design of a lightweight roof structure with a thermal system [31] • Using a digital twin to inform design choices for energy systems [32] • Storing thermal energy in buildings using digital twins [33]
4	Industry	• Industrial parallel intelligent control system (steam turbine and boiler systems in electricity stations, for example) [34]

Case Study: Rolls-Royce's "Intelligent Engine" Program

Rolls-Royce utilizes digital twins as part of its "Intelligent Engine" program, where each engine has its own digital twin. This technology enables real-time data collection from onboard sensors, monitoring over a dozen parameters during flights. As a result, the company can predict maintenance needs and reduce downtime. The primary goal of the program is to revolutionize how Rolls-Royce delivers power to its customers – by making engines more reliable and efficient and enhancing the overall customer experience.

c. *Automotive Industry*: Digital twins are pivotal in design, manufacturing, and aftermarket services. Engineers can experiment with multiple designs and identify the best-performing options without creating physical prototypes. This streamlines the design process while reducing costs. In manufacturing, digital twins help optimize production, reduce downtime, and boost overall efficiency. Additionally, predictive maintenance powered by digital twins enables the early detection of potential issues, allowing vehicles to be serviced before any problems occur. The advantages of digital twins in the automotive sector are vast, especially as the industry shifts toward electric vehicles (EVs). The need for digital twins in EV development and maintenance has become increasingly critical to ensure optimal performance, safety, and reliability.

Digital Twin Example: Volvo

Volvo, a leader in automobile safety, leverages digital twins to enhance its design process. The company can test different materials and aerodynamic properties for new vehicle designs by creating virtual replicas. This allows them to choose the most efficient designs, leading to improved performance and more fuel-efficient models without building multiple prototypes.

d. *Transportation Industry*: Digital twins are revolutionizing transportation by creating virtual models of roads, bridges, trains, and more. Engineers can test and optimize transportation systems before construction, simulating traffic congestion, weather conditions, and maintenance events. This leads to proactive maintenance and improved safety.

Digital Twin Example in Transportation: Ford

Ford employs digital twins in the development of autonomous vehicles. These virtual models allow Ford to test self-driving algorithms in various driving scenarios without needing physical prototypes. This reduces costs and speeds up development.

e. *Smart Cities and Urban Management*: Digital twins are also integral to the development of smart cities, offering a virtual environment to optimize urban planning, resource management, and sustainability efforts. By simulating scenarios such as traffic flow and emergency responses, city planners can enhance urban systems before making physical changes.

Digital Twin Example: Virtual Singapore Platform

The Virtual Singapore Platform uses a digital twin to replicate the city-state in great detail, helping planners address urban challenges such as infrastructure development and environmental management. This comprehensive virtual model allows for real-time data integration and accurate simulations, enabling better planning and decision-making for the city's future.

f. *Healthcare Industry*: In healthcare, digital twin technology enhances patient care, optimizes healthcare systems, and aids in training. Digital twins allow healthcare professionals to simulate different treatment scenarios, improving the accuracy of diagnoses and reducing unnecessary procedures [35–36]. Using digital twins, healthcare institutions can also model their operational features – such as patient flow and resource management. Hospitals can identify inefficiencies and implement solutions to improve efficiency by simulating various situations. For example, digital twins can assist with bed distribution and streamline treatment planning to ensure timely care for patients.

Digital Twin Example in Healthcare: Össur

Össur, a prosthetics and orthopedic solutions leader, uses digital twins to design custom-made prosthetics. This virtual modeling allows for highly personalized prosthetics, ensuring a better fit and functionality tailored to the needs of each individual.

g. *Digital Twin Technology across Industries: A Game-Changer*: The energy and utilities sector has significantly advanced its digital transformation in recent years, partly driven by the pandemic. One technology that has gained momentum is the digital twin. This technology enables companies to create virtual models of energy assets, such as power plants, to simulate real-time performance. The result? Reduced physical testing, lower costs, and optimized efficiency. Digital twins in energy allow businesses to build accurate virtual replicas of power generation systems, helping to simulate different operational scenarios. For example, by adjusting temperature or pressure, energy producers can optimize performance while minimizing machine wear and tear. Additionally, digital twins can improve the reliability of protection systems and enable remote testing, reducing the need for on-site evaluations.

Digital Twin Example in Energy: Shell

Shell uses digital twin technology for their offshore platforms. By collecting real-time sensor data, Shell can simulate various scenarios to optimize production and enhance safety. This approach helps them identify potential issues before they escalate and optimize operations. Beyond improving efficiency and safety, digital twins also play a role in worker training, offering realistic simulations to boost skill development.

13.4 FUTURE ENHANCEMENTS OF DIGITAL TWIN IN SMART INFRASTRUCTURE

Future developments in Digital Twin (DT) technology, especially in the context of Industry 5.0 and smart cities, hold transformative potential for emergency management and civil infrastructure (EMCI). With the integration of cutting-edge DT technologies into emergency management, this research explores how advancements in DT can enhance emergency preparedness, response, and recovery for civil infrastructure, as well as the future prospects of such applications. Through a systematic literature review, key research areas in the intersection of DT and EMCI were identified, including

(1) Advancements in DT development,
(2) The application of critical technologies, and
(3) The role of DT in enhancing infrastructure resilience and emergency management.

A proposed DT framework was developed to illustrate its applications at various stages of emergency management:

(1) Life cycle reinforcement during mitigation,
(2) Virtual planning, training, and decision-making for preparation,
(3) Real-time monitoring and optimization of resources in the response phase, and
(4) Collaborative learning during recovery efforts.

Several challenges and opportunities are anticipated for DT in emergency management for civil infrastructure. These include rapid modeling techniques that leverage vector and raster data with semantic enrichment, advancements in knowledge management to enable more dynamic and responsive emergency actions, and addressing cybersecurity and data-quality concerns through robust technical, contractual, and management solutions. This research offers essential insights for researchers and practitioners by systematically presenting knowledge on applying DTs in EMCI, promoting further development and innovation. The review highlights the early-stage nature of DTs in this field but stresses the urgency of establishing solid foundational concepts to ensure their effective implementation for societal benefit. The chapter also examines the role of DTs in Civil Infrastructure Systems (CIS) through literature reviews, surveys, and interviews with experts, revealing the substantial effort required from academia, industry, and government to realize DT potential fully. Based on the findings, the following recommendations are made:

1. **Redefining the Concept**: As a leading body in DT development, the Centre for Digital Built Britain (CDBB) should reconsider its definition of a DT. The definition proposed in this study should be shared for broader community input and refinement.
2. **Structured Implementation**: Collaboration is needed to systematize DT implementation processes, focusing on purpose-driven solutions. This will enhance understanding of the real-world asset, ensuring the digital twin's correct level of detail and accuracy.
3. **Quantifying and Visualizing Uncertainties**: Researchers should develop methods to quantify uncertainties in DT outputs and raise industry awareness on the importance of visualizing these uncertainties in decision-making.
4. **Focus on Short-Term and High-Value Use Cases**: DT research should prioritize applications with shorter time frames and those with higher safety or access risks, where precise and real-time monitoring is essential.

5. **Retrofitting Existing Infrastructure**: Further research is needed to explore how DTs can be integrated into existing infrastructure to improve operational efficiency and reduce energy usage.
6. **Data Sharing Frameworks**: Developing secure yet accessible frameworks for data sharing between interconnected DTs is critical to promoting collaboration without compromising data security.
7. **Digital Maturity Programs**: Organizations should invest in raising the digital maturity of their workforce before tackling more complex DT-related initiatives, ensuring smoother adoption and implementation.
8. **Centralized Oversight**: A dedicated organization should be established to oversee government-funded DT projects. This would promote a more coordinated approach, attract government investment, and ensure that DT developments serve public interests rather than purely corporate goals.

13.4.1 The Growing Importance of Digital Twins

From healthcare to energy, transportation, and beyond, digital twin technology is revolutionizing industries by providing virtual simulations of real-world scenarios. This capability fosters enhanced efficiency, safety, and innovation. As technology evolves, the adoption of digital twins is set to expand, offering even greater benefits to businesses, cities, and individuals. However, successful implementation hinges on addressing key challenges, including data privacy, security, and collaboration with specialized tech partners in digital twin solutions. By leveraging these powerful tools, industries can reduce costs, improve decision-making, and foster more sustainable, innovative futures. These efforts align with Industry 5.0's goal of building resilient, efficient, and smart infrastructure systems while addressing the increasing demands of emergency management. As digital twin technology advances – driven by progress in connectivity, artificial intelligence (AI), and machine learning (ML) – its widespread adoption faces challenges, particularly concerning data security and seamless integration. Originally developed for NASA's Apollo missions, digital twins have since become critical in sectors such as manufacturing, healthcare, and urban planning, demonstrating their transformative potential for efficiency and decision-making [37, 38, 39]. This shift reflects the principles of Industry 4.0, where sensor integration and advanced analytics play pivotal roles. However, as industries increasingly rely on digital twins, the importance of safeguarding sensitive data and adhering to regulatory standards grows. Striking a balance between harnessing data insights and ensuring robust security measures is essential [40, 41, 42]. To maintain the integrity of digital twin data, organizations must prioritize encryption, access controls, and secure communication protocols.

Solutions like Proqio address these concerns by offering a comprehensive platform that integrates IoT sensors and advanced data analytics for infrastructure projects. Proqio's secure and versatile system supports data management across all project phases, enabling organizations to optimize both security and efficiency in their digital twin applications. By adopting platforms like Proqio, businesses can foster a more resilient and integrated approach to digital twin technology within the Industry 4.0 framework [43].

13.5 CONCLUSION

Digital Twin (DT) technology is rapidly shaping the development of smart infrastructure by providing in-depth insights into the performance and longevity of urban systems. Through virtual replicas of physical assets, DTs facilitate ongoing monitoring, predictive analysis, and instant decision-making, ultimately enhancing project efficiency, safety, and sustainability. By capturing and analyzing extensive data from IoT sensors, DTs streamline maintenance, reduce operational costs, and improve resource allocation, allowing urban infrastructure to adapt to evolving needs. While the benefits are substantial, challenges such as data privacy, interoperability, and a lack of standardized practices must be addressed to unlock DT technology's full potential. As advances in AI, machine learning, and connectivity accelerate, DTs will play a pivotal role in smart city projects, aligning with the principles of Industry 4.0 and 5.0. By incorporating DTs, cities can build resilience, foster innovation, and support sustainable development, paving the way for safer and more adaptable urban environments for future generations.

REFERENCES

1. Evans, S., Savian, C., Burns, A., and Cooper, C., *Digital Twins for the Built Environment*, The IET (Institution of Engineering and Technology), London, 2020. Available at: https://www.theiet.org/impact-society/sectors/built-environment/ built-environment-news/2019-news/digital-twins-for-the-built-environment/ (accessed 12 May 2022).
2. Costello, K., and Omale, G., Gartner Survey Reveals Digital twins are Entering Mainstream Use, 2019. Available at: https://www.gartner.com/en/newsroom /press-releases/2019-02-20-gartner-survey-revealsdigital-twins-are-entering-mai (accessed 29 December 2020).
3. Enzer, M., Bolton, A., Boulton, C., Byles, D., Cook, A., Dobbs, L., El Hajj, P. A., Keaney, E., Kemp, A., Makri, C., Mistry, S., Mortier, R., Rock, S., Schooling, J., Scott, S., Sharp, M., West, M., and Winfield, M., *Roadmap for Delivering the Information Management Framework for the Built Environment*, CDBB, Cambridge, UK, 2019, https://doi.org/10.17863/CAM.38227.
4. Arup, Digital Twin – Towards a Meaningful Framework, 2018. Available at: https://www.arup.com/-/media/arup/files/publications/d/digital-twin-report. pdf (accessed 1 June 2020).
5. CDBB, Centre for Digital Built Britain, The Approach to Delivering a National Digital Twin for the United Kingdom, 2020. Available at: https://www.cdbb.cam .ac.uk/files/approach_summaryreport_final.pdf (accessed 1 June 2020).
6. Lamb, K., Principle-Based Digital Twins: A Scoping Review, 2019, Available at, https://www.cdbb.cam.ac.uk/files/scopingreview_dec20.pdf (accessed on 5 May 2020).
7. Teng, S., Chen, X., Chen, G., and Cheng, L., Structural damage detection based on transfer learning strategy using digital twins of bridges. *Mech. Syst. Signal Process.* 191, 110160, 2023. https://doi.org/10.1016/j.ymssp.2023.110160.

8. Liu, C., Xu, X., Wu, J., Zhu, H., and Wang, C., Deep transfer learning-based damage detection of composite structures by fusing monitoring data with physical mechanism. *Eng. Appl. Artif. Intell.* 123, 106245, 2023. https://doi.org/10.1016/j.engappai.2023.106245.
9. Liu, Z., Mohammed, O.A., and Liu, S., Equivalent hardware representation of PM synchronous motors from the physics-based phase variable model obtained through FE computation. *IEEE Trans. Magn.* 45(3), 1450–1453, 2009. https://doi.org/10.1109/TMAG.2009.2012670.
10. Lavanya Sharma et al., Impact of IoT on Big Data Analytics and Applications in Medical Images, Deep Learning in Internet of Things for Next Generation Healthcare Taylor & Francis, CRC Press, pp. 110–134, July 2024.
11. Lattner, A. D., Bogon, T., Lorion, Y., and Timm, I. J., A knowledge-based approach to automated simulation model adaptation. In: *Proceedings of the 2010 Spring Simulation Multiconference*, pp. 1–8, 2010. https://doi.org/10.1145/18785371878697 .
12. Wang, X., Wang, Y., Tao, F., and Liu, A., New paradigm of data-driven smart customisation through digital twin. *J. Manuf. Syst.* 58, 270–280, 2021. https://doi.org/10.1016/j.jmsy.2020.07.023.
13. Alexopoulos, K., Nikolakis, N., and Chryssolouris, G., Digital twin-driven supervised machine learning for the development of artificial intelligence applications in manufacturing. *Int. J. Comput. Integrated Manuf.* 33(5), 429–439, 2020. https://doi.org/10.1080/0951192X.2020.1747642.
14. Friederich, J., Francis, D. P., Lazarova-Molnar, S., Mohamed, N., A framework for data-driven digital twins for smart manufacturing. *Comput. Ind.* 136, 103586, 2022. https://doi.org/10.1016/j.compind.2021.103586.
15. Hosamo, H. H., and Hosamo, M. H., Digital Twin technology for bridge maintenance using 3D laser scanning: a review. *Adv. Civ. Eng.*, 2022(1), 2194949, 2022. https://doi.org/10.1155/2022/2194949.
16. Mill, T., Alt, A., and Liias, R., Combined 3D building surveying techniques–terrestrial laser scanning (TLS) and total station surveying for BIM data management purposes. *J. Civ. Eng. Manag.* 19(Suppl. 1), S23–S32, 2013. https://doi.org/10.3846/13923730.2013.795187.
17. Mohammadi, M., Rashidi, M., Mousavi, V., Karami, A., Yu, Y., and Samali, B., Quality evaluation of digital twins generated based on UAV photogrammetry and TLS: bridge case study. *Rem. Sens.* 13(17), 3499, 2021. https://doi.org/10.3390/rs13173499.
18. Wang, S., Rodgers, C., Zhai, G., Matiki, T. N., Welsh, B., Najafi, A., Wang, J., Narazaki, Y., Hoskere, V., and Spencer Jr., B. F., A graphics-based digital twin framework for computer vision-based post-earthquake structural inspection and evaluation using unmanned aerial vehicles. *J. Infrastruct. Intell Resilience* 1(1), 100003, 2022. https://doi.org/10.1016/j.iintel.2022.100003.
19. Bolton, A., Enzer, M., and Schooling, J., *The Gemini Principles: Guiding Values for the National Digital Twin and Information Management Framework*, Centre for Digital Built Britain and Digital Framework Task Group, Cambridge, UK, 2018, https://doi.org/10.17863/CAM.32260. Available at: https://www.cdbb.cam.ac. uk/system/files/documents/TheGeminiPrinciples.pdf (accessed 9 February 2022).
20. Ding, W., Yang, H., Yu, K., and Shu, J., Crack detection and quantification for concrete structures using UAV and transformer. *Autom. Constr.* 152, 104929, 2023. https://doi.org/10.1016/j.autcon.2023.104929.
21. Levine, N. M., and Spencer Jr., B. F., Post-earthquake building evaluation using UAVs: aBIM-based digital twin framework. *Sensors* 22(3), 873, 2022. https://doi.org/10.3390/s22030873.

22. Teng, S., Chen, X., Chen, G., and Cheng, L., Structural damage detection based on transfer learning strategy using digital twins of bridges. *Mech. Syst. Signal Process.* 191, 110160, 2023. https://doi.org/10.1016/j.ymssp.2023.110160.
23. Wang, S., Rodgers, C., Zhai, G., Matiki, T. N., Welsh, B., Najafi, A., Wang, J., Narazaki, Y., Hoskere, V., and Spencer Jr., B. F., A graphics-based digital twin framework for computer vision-based post-earthquake structural inspection and evaluation using unmanned aerial vehicles. *J. Infrastruct. Intelligence Resilience* 1(1), 100003, 2022. https://doi.org/10.1016/j.iintel.2022.100003
24. Shu, J., Zhang, C., Chen, X., and Niu, Y., Model-informed deep learning strategy with vision measurement for damage identification of truss structures. *Mech. Syst. Signal Process.* 196, 110327, 2023. https://doi.org/10.1016/j.ymssp.2023.110327
25. Borjigin, A. O., Sresakoolchai, J., Kaewunruen, S., and Hammond, J., Digital twin aided sustainability assessment of modern light rail infrastructures. *Front. Built Environ.* 8, 796388, 2022. https://doi.org/10.3389/fbuil.2022.796388.
26. Kaewunruen, S., and Xu, N., Digital twin for sustainability evaluation of railway station buildings. *Front. Built Environ.* 4, 77, 2018. https://doi.org/10.3389/fbuil.2018.00077 .
27. Lavanya Sharma et al., Deep Learning Processes in MRI Images, Deep Learning in Internet of Things for Next Generation Healthcare Taylor & Francis, CRC Press, pp. 210–229, July 2024.
28. Liu, F., Xun, J., Zhou, M., He, S., and Dong, H., October. A driving strategy based integrated rescheduling model for high-speed railway by using the parallel intelligent method. In: *2022 IEEE 2nd International Conference on Digital Twins and Parallel Intelligence (DTPI)*, pp. 1–6. IEEE, 2022. https://doi.org/10.1109/DTPI55838.2022.9998978.
29. Seo, H., Yun, and W.-S., Digital twin-based assessment framework for energy savings in university classroom lighting. *Buildings* 12 (5), 544, 2022. https://doi.org/10.3390/buildings12050544.
30. Sharma, L., & Garg, P.K. (Eds.). (2024). Deep Learning in Internet of Things for Next Generation Healthcare (1st ed.). Chapman and Hall/CRC. https://doi.org/10.1201/9781003451846
31. Lydon, G. P., Caranovic, S., Hischier, I., and Schlueter, A., Coupled simulation of thermally active building systems to support a digital twin. *Energy Build.* 202, 109298, 2019. https://doi.org/10.1016/j.enbuild.2019.07.01.
32. Granacher, J., Nguyen, T.-V., Castro-Amoedo, R., and Mar_echal, F., Overcoming decision paralysis—a digital twin for decision making in energy system design. *Appl. Energy* 306, 117954, 2022. https://doi.org/10.1016/j.apenergy.2021.117954.
33. Lv, Z., Li, Y., Feng, H., and Lv, H., Deep learning for security in digital twins of cooperative intelligent transportation systems. *IEEE Trans. Intell. Transport. Syst.* 23 (9), 16666–16675, 2021. https://doi.org/10.1109/TITS.2021.3113779.
34. Li, Q., Su, Y., Yin, F., Dong, L., Sun, J., and Zhang, X., A parallel intelligent control system and its industrial application. In: *2021 IEEE 1st International Conference on Digital Twins and Parallel Intelligence (DTPI)*, pp. 9–12. IEEE, October 2021. https://doi.org/10.1109/DTPI52967.2021.9540189.
35. Hooda, Y., and Singh, H.D., Advancement of Geospatial Technology in Healthcare Systems. In *Deep Learning in Internet of Things for Next Generation Healthcare* (pp. 76–87). Chapman and Hall/CRC, 2024.
36. Singh, H. D., Hooda, Y., and Mehta, J., Disaster and Emergency Healthcare. In *Deep Learning in Internet of Things for Next Generation Healthcare* (pp. 120–127). Chapman and Hall/CRC, 2024.

37. Hooda, Y., and Singh, H. D., Digital Reforms in Public Services and Infrastructure Development & Management. In *Technological Prospects and Social Applications of Society 5.0* (pp. 219–237). Chapman and Hall/CRC, 2023.
38. Hooda, Y., IoT and remote sensing. In *Computer Vision and Internet of Things* (pp.111–140). Chapman and Hall/CRC, 2022.
39. Sharma, L., & Garg, P.K. (Eds.). (2023). Technological Prospects and Social Applications of Society 5.0 (1st ed.). Chapman and Hall/CRC. https://doi.org/10.1201/9781003324720 *[Indexed in Scopus]*
40. Sharma, L., & Carpenter, M. (Eds.). (2022). *Computer Vision and Internet of Things: Technologies and Applications* (1st ed.). Chapman and Hall/CRC. https://doi.org/10.1201/9781003244165
41. Sharma, L., & Garg, P.K. (Eds.). (2021). *Artificial Intelligence: Technologies, Applications, and Challenges* (1st ed.). Chapman and Hall/CRC. https://doi.org/10.1201/9781003140351
42. Sharma, L. (Ed.). (2020). Towards Smart World: Homes to Cities Using Internet of Things (1st ed.). Chapman and Hall/CRC. https://doi.org/10.1201/9781003056751
43. Sharma, L., & Garg, P.K. (Eds.). (2019). From Visual Surveillance to Internet of Things: Technology and Applications (1st ed.). Chapman and Hall/CRC. https://doi.org/10.1201/9780429297922

14 Navigating the Complexities

Limitations and Challenges of Digital Twin Technology in Smart Cities

Sudesh Agrawal, Naman Agrawal, Himanshu Agrawal, and Sonal Agrawal

14.1 INTRODUCTION: THE PROMISE AND POTENTIAL OF DIGITAL TWINS

Digital Twin technology has emerged as one of the most transformative innovations in urban planning and infrastructure management. By creating a dynamic, real-time virtual model of a physical entity, a Digital Twin allows city planners to simulate, monitor, and optimize urban systems with unparalleled accuracy. As cities worldwide face the challenges of rapid urbanization, climate change, and resource constraints, the use of Digital Twins offers the potential to improve sustainability, resilience, and efficiency in smart city environments. The global market for Digital Twin technology is projected to grow from USD 3.1 billion in 2020 to USD 48.2 billion by 2026, highlighting its increasing adoption across various sectors (MarketsandMarkets, 2020).

At its core, a Digital Twin integrates real-time data from Internet of Things (IoT) sensors, historical data, and advanced analytics to create a virtual replica of physical infrastructure, such as buildings, transportation networks, energy grids, and even entire cities. This real-time simulation enables cities to monitor and predict how changes in one part of the system affect the whole. For instance, Amsterdam's Digital Twin has helped the city reduce energy consumption by 18% in public buildings and improved traffic flow by 12% through real-time simulations of traffic patterns (City of Amsterdam, 2021a).

Urban planners and municipal governments are increasingly turning to Digital Twins to address the complexities of modern cities. The technology allows cities to simulate potential impacts of new policies or urban developments before they are implemented. For example, Singapore's Virtual Singapore initiative, a nationwide

DOI: 10.1201/9781003582489-14

Digital Twin project, allows the government to simulate urban scenarios like new construction, disaster management, and transportation planning (Government Technology Agency, 2021). This predictive capability provides significant cost savings and improves decision-making processes by identifying inefficiencies before they manifest in real life.

The potential of Digital Twins to drive sustainability is also significant. By enabling cities to optimize energy usage, manage traffic more effectively, and predict infrastructure wear and tear, Digital Twins can play a critical role in helping cities meet their sustainability goals. For instance, Rotterdam's Digital Twin Port has optimized logistics and energy use, reducing carbon emissions by 15% while also improving port efficiency by 20% (Port of Rotterdam Authority, 2022). These improvements are not only vital for meeting environmental targets but also for ensuring cities can manage their growing populations without overwhelming their infrastructure.

However, despite the promise of Digital Twins, scaling these models across cities and industries presents a set of complex challenges. Integrating diverse data sources – such as IoT sensors, Geographic Information Systems (GIS), and traffic management systems – into a unified Digital Twin platform is a technical feat that many cities are struggling to achieve. Additionally, developing nations often lack the infrastructure investment required to implement these technologies at scale (McKinsey & Company, 2021). Moreover, cybersecurity vulnerabilities and data privacy concerns add further layers of complexity to Digital Twin implementation, especially when real-time data from critical infrastructure is involved (RICS, 2021; Figure 14.1).

In summary, while the potential of Digital Twin technology to revolutionize urban planning is immense, it is equally important to address the limitations and challenges that come with it. As more cities embrace this technology, they must also navigate the complexities of data integration, cybersecurity, infrastructure costs, and ethical considerations to fully realize the transformative power of Digital Twins in creating smarter, more sustainable cities.

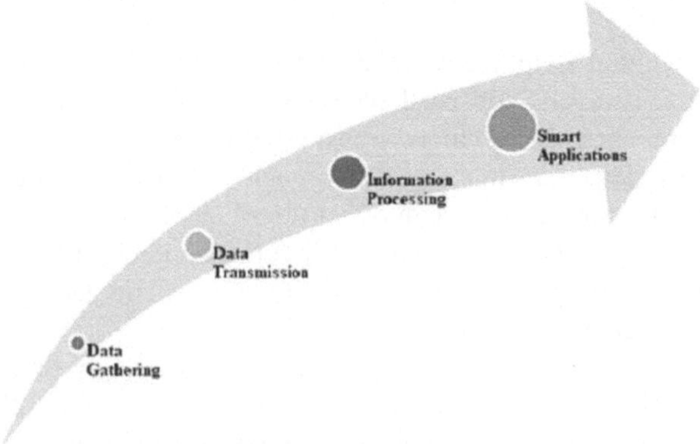

FIGURE 14.1 The fundamental architecture of Internet of Things (IoT)

14.2 DATA INTEGRATION: BRIDGING THE GAPS BETWEEN SYSTEMS

One of the most significant challenges in deploying Digital Twin technology for smart city applications is data integration. Digital Twins rely on vast amounts of real-time data from diverse sources to create accurate virtual representations of physical environments. However, integrating data from various systems – such as IoT sensors, Geographic Information Systems (GIS), traffic management systems, and public utilities – into a unified platform is no small feat. As cities expand and adopt more complex systems, the challenge of data interoperability, integration, and harmonization becomes increasingly pressing (sharma, l., 2020).

For Digital Twin technology to realize its full potential, data must flow seamlessly between numerous and often siloed systems. Yet, many cities and municipalities find themselves grappling with legacy systems that are not designed to interact with newer, smarter technologies. Moreover, the costs associated with upgrading infrastructure to handle the enormous data loads required for Digital Twins often prove prohibitive, particularly in developing regions. This section delves into the key challenges of data integration for Digital Twin models and explores solutions to bridge these gaps effectively.

At the heart of a successful Digital Twin implementation is its ability to collect and process real-time data from a wide array of sources. For example, a citywide Digital Twin might rely on data from IoT sensors embedded in roads to monitor traffic flow, from weather stations to predict environmental impacts, and from utility systems to track energy usage. Each of these systems generates massive amounts of data in different formats, using various communication protocols. McKinsey & Company (2021) notes that the challenge of managing such diverse data streams is one of the primary obstacles to widespread Digital Twin adoption (sharma.L., 2022).

One key example is Singapore's Virtual Singapore project, where the city's Digital Twin integrates data from over 1,000 different data sources, including transportation, urban planning, public safety, and environmental monitoring systems. Despite the city's cutting-edge infrastructure, data interoperability remains a significant hurdle. The sheer volume of data, often collected in incompatible formats, requires advanced data harmonization tools to ensure that it can be used effectively in simulations and decision-making processes (Government Technology Agency, 2021). For cities with less advanced infrastructure, such integration challenges become even more pronounced.

In many cases, legacy systems – such as outdated traffic management networks or older utility grids – are not compatible with newer IoT-based technologies. Cities like New York, for example, have struggled to integrate older traffic light systems with AI-powered Digital Twin platforms. This lack of system compatibility hinders the ability to provide real-time traffic optimizations that could reduce congestion and improve urban mobility (NYC Mayor's Office of Technology, 2020a, e).

14.2.1 The Role of Data Standards and Interoperability

A major issue contributing to data integration challenges is the absence of standardized protocols for data collection and sharing between different systems. Each city department or service provider might use different data formats, making it difficult to combine information into a comprehensive Digital Twin. For instance, traffic sensors may use one communication protocol, while utility management systems might use another, leading to data silos that hinder effective coordination.

To address these challenges, many cities and organizations are calling for the adoption of open data standards and frameworks. The Open Geospatial Consortium (OGC), for example, has developed standards for geospatial data sharing that are now being used by cities to improve data interoperability. By adhering to these standards, cities can ensure that data from GIS systems, environmental sensors, and utility networks can be integrated more easily into a unified Digital Twin model (OGC, 2021).

Barcelona, a global leader in smart city initiatives, has made significant strides in standardizing its data collection practices improving interoperability across systems. The city's Digital Twin integrates data from its smart waste management system, public transportation networks, and energy grids by using open standards that allow these systems to communicate seamlessly. This approach has enabled Barcelona to achieve a 20% improvement in energy efficiency and a 15% reduction in operational costs for public services (City of Barcelona, 2022).

14.2.2 Cloud Computing and Edge Computing: Enablers of Data Integration

Given the vast amounts of data generated by Digital Twins, cloud computing has emerged as a critical tool for enabling effective data integration. Cloud platforms allow cities to store, process, and analyze data from multiple sources in real time, providing the computational power needed to manage the complexity of urban environments. Leading cloud providers such as Amazon Web Services (AWS) and Microsoft Azure are increasingly offering specialized services for Digital Twin development, allowing cities to leverage scalable infrastructure for their smart city initiatives (Amazon Web Services, 2021).

In addition to cloud computing, edge computing is becoming a key enabler of data integration for Digital Twins. Unlike cloud computing, which relies on centralized data processing, edge computing allows data to be processed closer to where it is generated. This approach reduces latency and improves real-time decision-making, making it especially useful for applications such as traffic management and public safety. Copenhagen, for example, uses edge computing in its Digital Twin model to process data from traffic cameras and environmental sensors in real time, allowing the city to adjust traffic signals and reduce congestion without the need for cloud-based processing (City of Copenhagen, 2022a, c).

By combining cloud and edge computing, cities can achieve a more flexible and efficient approach to data integration, enabling Digital Twins to function effectively in even the most complex urban environments. However, for many cities, the cost of implementing such infrastructure remains a significant barrier.

14.2.3 Data Integration in Developing Cities: A Costly Challenge

While cities in developed nations are making strides in data integration, many developing cities face substantial challenges in building the necessary infrastructure to support Digital Twin technology. The costs associated with upgrading legacy systems, installing IoT sensors, and establishing cloud or edge computing capabilities can be prohibitive, particularly for cities with limited budgets.

For instance, Mumbai, one of India's most populous cities, has begun exploring Digital Twin technology to improve its traffic management and public safety systems. However, the city's existing infrastructure is not equipped to handle the massive data loads required for a comprehensive Digital Twin model. According to McKinsey & Company (2021), the cost of upgrading Mumbai's data infrastructure to support Digital Twin applications would run into hundreds of millions of dollars, a significant hurdle for the city's municipal government.

To bridge this gap, international collaborations and public–private partnerships are emerging as critical strategies. The Indo-Dutch partnership, for example, is working to implement Digital Twin technology in cities like Hyderabad, where Dutch expertise in data integration and urban mobility is helping the city overcome its infrastructural limitations. This collaboration has allowed Hyderabad to integrate its traffic management system with real-time IoT data, leading to a 25% reduction in traffic congestion in pilot zones (IIT Hyderabad, 2023). Such partnerships can serve as a model for other developing cities facing similar challenges.

14.2.4 The Path Forward: Best Practices for Data Integration

As cities continue to adopt Digital Twin technology, several best practices have emerged for overcoming data integration challenges (sharma L., 2022). These include:

1. Adopting Open Standards: Cities should implement open data standards to ensure that different systems can communicate effectively. This requires collaboration between governments, technology providers, and international organizations to establish common protocols for data sharing.
2. Investing in Cloud and Edge Computing: Leveraging cloud and edge computing solutions can help cities manage the enormous data loads generated by Digital Twins while improving real-time decision-making.
3. Building Public–Private Partnerships: Collaborating with private companies and international organizations can provide the technical expertise and funding needed to upgrade infrastructure and integrate diverse data systems.

4. Gradual Implementation: Cities with limited resources should consider a phased approach to Digital Twin implementation, starting with smaller pilot projects before scaling up to citywide initiatives.

By addressing the challenges of data integration, cities can unlock the full potential of Digital Twin technology, creating smarter, more efficient, and sustainable urban environments.

14.3 INFRASTRUCTURE REQUIREMENTS: THE COST OF BUILDING DIGITAL TWINS

While Digital Twin technology promises to revolutionize urban planning and infrastructure management, the journey to fully operational Digital Twins is resource-intensive and demands significant infrastructure investment. For a Digital Twin to function optimally, cities must invest in robust data collection mechanisms, scalable computing resources, and advanced connectivity systems such as 5G. These infrastructure components underpin the seamless flow of data necessary for Digital Twins to provide real-time simulations, predictive analytics, and valuable insights. However, the cost of building and maintaining this infrastructure can be prohibitive, particularly for developing cities with limited financial and technological resources.

This section explores the major infrastructure requirements for Digital Twins, the associated costs, and how cities around the world are managing these challenges. It also highlights strategies for overcoming cost barriers, including public–private partnerships, cloud-based solutions, and international collaboration.

14.3.1 IoT Sensors: The Backbone of Real-Time Data Collection

The first critical infrastructure component of a Digital Twin is IoT sensors, which collect real-time data from the physical environment. These sensors, embedded in roads, buildings, utility systems, and public spaces, serve as the eyes and ears of the Digital Twin, feeding it with continuous streams of data on everything from traffic flow to energy consumption.

For a comprehensive Digital Twin to function effectively, thousands of IoT sensors are required. According to McKinsey & Company (2021), the average smart city relies on at least 200 IoT sensors per square kilometer to collect sufficient data. The cost of deploying these sensors, combined with the necessary communication networks to transmit the data, can be a significant expense for cities. In Singapore, for example, the Virtual Singapore project, which aims to create a Digital Twin of the entire nation, has invested heavily in IoT infrastructure, with costs exceeding USD 73 million for the installation of IoT sensors and associated systems (Government Technology Agency, 2021).

Developing cities face even greater challenges when it comes to the cost of IoT deployment. In Nairobi, Kenya, the local government began piloting Digital Twin technology to manage traffic congestion and optimize energy consumption. However, the lack of existing IoT infrastructure meant that Nairobi had to invest in building its data collection systems from scratch, costing an estimated USD 10 million for the installation of IoT sensors in just the city center (McKinsey & Company, 2021).

14.3.2 Cloud Computing and Data Storage: Handling Massive Data Volumes

Once data is collected, it must be processed, stored, and analyzed tasks that require vast computational power and storage capacity. Cloud computing has emerged as a vital enabler of Digital Twin technology, offering cities the ability to scale their data processing needs without investing in expensive, on-premises hardware. Cloud platforms, such as Amazon Web Services (AWS) and Microsoft Azure, provide flexible infrastructure solutions that allow cities to handle the large data volumes generated by Digital Twins.

However, the cost of cloud services can quickly escalate as the amount of data being processed grows. According to Gartner (2022), cities using Digital Twin technology may need to process up to 10 terabytes of data per day, depending on the complexity of their systems. For smaller cities or developing nations, this presents a financial challenge, as cloud services can cost upwards of USD 1 million per year for advanced Digital Twin applications (Gartner, 2022).

Despite these costs, many cities have embraced cloud-based solutions as a cost-effective way to manage the high computational demands of Digital Twins. Amsterdam's Digital Twin City Project, for example, relies on Microsoft Azure's cloud infrastructure to process data from over 1,200 IoT sensors across the city. By leveraging cloud services, Amsterdam has been able to scale its Digital Twin without incurring the high costs associated with maintaining its own data centers (City of Amsterdam, 2021c, d).

14.3.3 High-Speed Connectivity: Enabling Real-Time Simulations

Another key infrastructure requirement for Digital Twins is high-speed connectivity, particularly as cities move toward real-time data processing and simulation. 5G networks are seen as essential for enabling the low-latency communication needed for Digital Twins to function effectively. With 5G, data can be transmitted between IoT sensors and central processing units almost instantaneously, allowing Digital Twins to make real-time adjustments to urban systems like traffic lights, public transportation, and energy grids.

The rollout of 5G, however, comes with significant costs. According to a report by Deloitte (2021), building a 5G network in a large city costs between USD 1 billion and USD 10 billion, depending on the scope of the deployment. For developing nations, these costs can be particularly prohibitive. For example, in India, where the government is eager to implement 5G to support the country's Smart Cities Mission, initial estimates suggest that the cost of 5G deployment across the country could exceed USD 30 billion (Deloitte, 2021).

Nevertheless, cities that have invested in 5G are already seeing the benefits. South Korea's capital, Seoul, has implemented 5G-enabled Digital Twins to optimize its public transport network. By using 5G to process real-time data on passenger flow and traffic conditions, Seoul has been able to reduce congestion by 20% on key bus routes (Kim, 2022). These examples demonstrate that while 5G deployment is expensive, the long-term benefits in terms of efficiency and sustainability can outweigh the initial costs.

14.3.4 The Challenge of Legacy Infrastructure

In addition to the financial cost of building new infrastructure, many cities must also contend with the challenge of legacy systems. Older infrastructure – such as traffic management systems, energy grids, and public utilities – often lacks the capability to integrate with modern IoT devices or cloud platforms. Upgrading or replacing these legacy systems can be a costly and time-consuming process, but it is essential for enabling the full potential of Digital Twin technology.

In New York City, for instance, the Department of Transportation faced significant challenges when attempting to integrate its outdated traffic light system with a new Digital Twin platform designed to reduce congestion. The cost of upgrading New York's 13,000 traffic lights to be compatible with IoT sensors and real-time analytics systems was estimated at USD 500 million (NYC Mayor's Office of Technology, 2020a, b). Similarly, London's Underground system required significant investment to modernize its signaling infrastructure before it could be integrated into the city's Digital Twin for Transport project, which cost over GBP 1 billion (Transport for London, 2021).

14.3.5 Strategies for Overcoming Cost Barriers

Despite the high costs associated with building the infrastructure for Digital Twins, several strategies are being employed by cities to overcome these financial barriers. One such strategy is the development of public–private partnerships (PPPs), which allow cities to share the cost of infrastructure development with private companies. Barcelona, for example, partnered with private tech firms to finance its smart city initiatives, allowing the city to implement IoT sensors and cloud-based solutions without bearing the full cost (City of Barcelona, 2022).

Another solution is the use of cloud-based Digital Twin platforms that offer a more flexible and scalable alternative to on-premises infrastructure. Cities like Amsterdam and Copenhagen have adopted cloud solutions to reduce the upfront investment in physical infrastructure, allowing them to scale their Digital Twin systems as needed without significant capital expenditures United Nations (2021).

Additionally, international collaboration can provide financial and technical support to cities in developing nations. The Indo-Dutch partnership in Hyderabad, for example, has provided the city with access to Dutch expertise and funding, helping to reduce the cost of implementing Digital Twin technology for traffic management and urban mobility (IIT Hyderabad, 2023).

14.4 CYBERSECURITY THREATS: THE VULNERABILITIES OF DIGITAL TWINS

As Digital Twin technology becomes increasingly integrated into smart city infrastructure, the issue of cybersecurity has emerged as a critical concern. By creating a virtual replica of physical systems and processes, Digital Twins rely on continuous streams of real-time data from IoT devices, sensors, and other connected systems. While this integration offers unprecedented insights into urban management, it also opens the door to a host of cybersecurity vulnerabilities. The potential for cyberattacks on Digital Twins poses significant risks, from disrupting essential services like energy grids and public transportation to endangering public safety and privacy. As more cities adopt Digital Twin technology, ensuring the security of these systems has become a top priority.

This section delves into the cybersecurity threats faced by Digital Twin implementations, examines real-world examples of cyberattacks on critical infrastructure, and explores strategies for mitigating these risks. In a world where smart cities are increasingly reliant on interconnected systems, addressing cybersecurity vulnerabilities is essential to realizing the full potential of Digital Twin technology.

14.4.1 The Nature of Cybersecurity Threats in Digital Twin Environments

The complexity of Digital Twin systems makes them an attractive target for cybercriminals. These systems collect vast amounts of data from various sources, including critical infrastructure such as power plants, water treatment facilities, public transport systems, and healthcare networks. The sheer scale and diversity of the data involved create multiple points of vulnerability. A successful cyberattack on a Digital Twin could lead to unauthorized access to sensitive information, system manipulation, or even complete control over physical assets.

McKinsey & Company (2021) notes that as the use of IoT devices in smart cities continues to expand, so too does the attack surface for potential cyber intrusions. IoT sensors and devices are often the weakest link in a Digital Twin system, as they are distributed across urban environments and frequently lack robust security protocols. A single compromised IoT device can provide a gateway for attackers to access the entire Digital Twin system, causing widespread disruption.

Moreover, the real-time nature of Digital Twins adds another layer of complexity to cybersecurity. Unlike traditional static systems, which can be more easily monitored and updated, Digital Twins rely on continuous data streams. This makes it difficult to detect and mitigate threats in real time. In many cases, cyberattacks on Digital Twins are designed to remain undetected for extended periods, allowing attackers to exploit vulnerabilities without triggering immediate alarms (Figure 14.2).

14.4.2 Real-World Examples of Cyberattacks on Critical Infrastructure

Several high-profile cyberattacks on critical infrastructure in recent years highlight the vulnerabilities inherent in systems that rely on real-time data and interconnected networks – features that are central to Digital Twins.

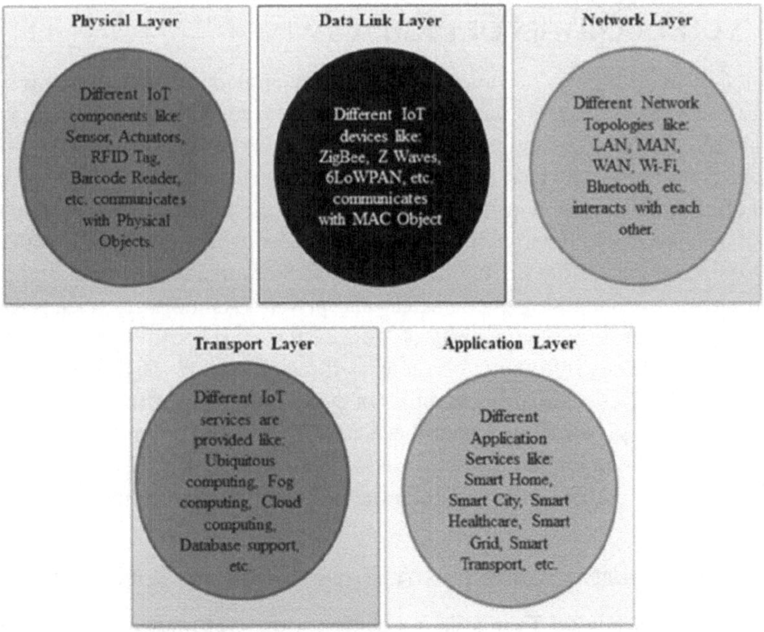

FIGURE 14.2 Different layers of Internet of Things (IoT) Services

One of the most notable examples is the 2019 cyberattack on a US water treatment plant, where hackers gained access to the plant's control systems through a compromised IoT device. The attackers were able to manipulate the chemical levels in the water supply, posing a serious threat to public safety (RICS, 2021). While the attack was eventually thwarted before causing harm, it exposed the potential dangers of leaving IoT-based critical infrastructure vulnerable to cyberattacks.

Similarly, in Ukraine, a 2016 cyberattack on the country's power grid led to widespread blackouts affecting more than 230,000 residents (Zetter, 2017). The attackers gained access to the grid's control systems through malicious phishing emails and malware, allowing them to remotely disable power substations. Although this incident did not directly involve a Digital Twin, it underscores the risks of cyberattacks on critical infrastructure systems that could be integrated into Digital Twins.

More recently, the 2021 ransomware attack on the Colonial Pipeline in the U.S. highlighted the economic and security implications of a cyberattack on critical energy infrastructure. Hackers gained access to the pipeline's operational technology network, forcing a shutdown of the pipeline, and leading to fuel shortages across the southeastern United States (CISA, 2021). As cities begin to use Digital Twins to manage energy grids and other vital infrastructure, similar attacks could have even more devastating consequences.

14.4.3 The Threat of Data Manipulation and System Sabotage

One of the most significant risks posed by cyberattacks on Digital Twins is data manipulation. Because Digital Twins rely on real-time data to simulate and optimize physical systems, altering this data could lead to dangerous or costly decisions. For example, if an attacker were to manipulate traffic data in a city's Digital Twin, it could lead to traffic signal failures or accidents. Similarly, manipulating data from energy grids could cause outages or overloads that damage infrastructure and disrupt essential services.

Additionally, Digital Twins are increasingly being used to support decision-making in disaster response and urban planning. A compromised Digital Twin could provide inaccurate or misleading information, leading to ineffective or harmful responses during emergencies. The ability to sabotage a city's infrastructure through data manipulation represents a significant security risk that cities must address as they implement Digital Twin technology.

In Singapore, the government has invested heavily in cybersecurity measures to protect its Virtual Singapore project from these kinds of threats. The city-state's Cyber Security Agency (CSA) has implemented rigorous security protocols, including end-to-end encryption for data transmission, multi-factor authentication, and AI-based threat detection systems (Government Technology Agency, 2021). These measures are designed to prevent unauthorized access to the Digital Twin system and ensure that data integrity is maintained.

14.4.4 Privacy Concerns and the Ethics of Surveillance

Beyond the risk of service disruption or infrastructure sabotage, privacy and surveillance concerns present another cybersecurity challenge for Digital Twin technology. Many Digital Twins rely on data from citizens' devices, such as smartphones, GPS systems, and even social media platforms, to model human behavior and improve public services. While this data can provide valuable insights into traffic patterns, energy usage, and public health trends, it also raises ethical questions about how much personal data cities should collect and how that data is protected.

For instance, in China, several cities have implemented Digital Twins as part of their broader smart city initiatives, which use surveillance cameras and facial recognition technology to monitor citizens in real time. While this data is used to optimize urban systems, it has also been criticized for enabling mass surveillance and infringing on privacy rights (Wang, 2021). Ensuring that Digital Twin technology is deployed ethically, with robust privacy protections in place, is essential to maintaining public trust.

14.4.5 Solutions for Securing Digital Twins

Given the range of cybersecurity threats facing Digital Twin technology, cities must adopt a comprehensive approach to securing their systems. Some of the most effective strategies include:

1. End-to-End Encryption: Ensuring that data is encrypted from the moment it is collected by IoT sensors until it is processed in the Digital Twin system is critical for preventing unauthorized access. Encryption helps to protect sensitive information, making it much more difficult for attackers to intercept or alter data.
2. Zero Trust Security Model: The Zero Trust model assumes that no device or user, whether inside or outside the network, should be trusted by default. This approach requires every user or device to be authenticated, authorized, and continuously validated before gaining access to the system. Implementing Zero Trust security for Digital Twins can reduce the risk of unauthorized access (Forrester, 2021).
3. AI-Powered Threat Detection: Leveraging artificial intelligence (AI) to monitor Digital Twin systems for unusual activity or anomalies can help cities detect cyber threats in real time. AI algorithms can analyze data patterns to identify potential breaches before they cause significant damage. For example, Copenhagen's Digital Twin uses AI-based security tools to continuously monitor for cybersecurity threats across its public transportation network (City of Copenhagen, 2022a).
4. Regular Security Audits: Conducting regular cybersecurity audits of the Digital Twin system can help identify vulnerabilities and ensure that security protocols are up to date. These audits should include penetration testing, where cybersecurity experts simulate attacks on the system to uncover weaknesses.
5. Public Awareness and Ethical Guidelines: To address privacy concerns, cities should adopt clear ethical guidelines on the collection and use of personal data. These guidelines should be transparent, allowing citizens to understand how their data is being used and ensuring that privacy is protected.

As cities continue to embrace Digital Twin technology to enhance urban management, they must also recognize the significant cybersecurity risks associated with these systems. The integration of real-time data from IoT devices and critical infrastructure creates multiple points of vulnerability that cybercriminals can exploit. From data manipulation to infrastructure sabotage, the consequences of a successful cyberattack on a Digital Twin could be catastrophic.

However, by implementing robust security measures – such as end-to-end encryption, Zero Trust security models, and AI-powered threat detection – cities can mitigate these risks and ensure the safety and reliability of their Digital Twin systems. In addition, addressing privacy concerns through ethical data usage guidelines is essential for maintaining public trust. As the adoption of Digital Twins continues to expand, a proactive approach to cybersecurity will be critical to unlocking the full potential of this transformative technology.

14.5 REGULATORY CHALLENGES: GOVERNANCE AND STANDARDIZATION

The rapid adoption of Digital Twin technology in smart cities presents significant regulatory challenges related to governance and standardization. As cities increasingly rely on Digital Twins to optimize urban infrastructure, manage resources, and improve public services, the lack of clear regulatory frameworks and standardized protocols has emerged as a critical issue. Without proper governance, cities may struggle to manage the complexities of data ownership, privacy, and interoperability, leading to inefficiencies and potential misuse of technology.

One of the primary regulatory challenges is the absence of global standards for Digital Twin implementation. Different cities and countries are developing their own approaches to Digital Twin technology, leading to fragmentation in how data is collected, processed, and shared. For instance, Amsterdam and Singapore have adopted different data protocols for their Digital Twin systems, making it difficult for cities to collaborate or share best practices (City of Amsterdam, 2021b, d; Government Technology Agency, 2021). This lack of standardization hinders the potential for cross-border collaboration and prevents the development of scalable solutions that can be implemented globally.

The issue of data governance is also central to the regulatory debate. Digital Twins depend on vast amounts of real-time data, much of which may be personal or sensitive. As cities collect data from IoT devices, sensors, and personal devices, questions arise about who owns this data, how it should be protected, and how it can be used ethically. The European Union's General Data Protection Regulation (GDPR) provides a framework for data privacy and ownership in Europe, but many countries lack equivalent protections, leading to a regulatory vacuum in regions where Digital Twins are being rapidly deployed (European Commission, 2021).

Moreover, the challenge of interoperability requires global standards to ensure that Digital Twin systems can communicate with one another effectively. The Open Geospatial Consortium (OGC) and the International Organization for Standardization (ISO) have been working on developing smart city standards, but progress has been slow. Without universal standards, cities may become dependent on specific technology providers, leading to vendor lock-in and limiting the flexibility to expand or modify Digital Twin systems over time (ISO, 2020a, b).

Addressing these regulatory challenges requires global cooperation and the development of clear frameworks for governance, data privacy, and interoperability. By establishing universal standards, cities can ensure that Digital Twin technology is implemented in a way that protects citizens' rights and enables seamless collaboration between urban centers World Bank (2020).

14.6 CONCLUSION AND WAY FORWARD

Despite the significant challenges faced in the adoption and implementation of Digital Twin technology, there is a clear path forward to overcome these obstacles and unlock the full potential of this transformative technology in smart cities. By addressing issues related to cost, infrastructure, cybersecurity, and regulation, cities can successfully integrate Digital Twins to create more efficient, sustainable, and resilient urban environments.

One of the most promising approaches to overcoming the challenges of cost and scalability is the use of cloud-based and edge computing solutions. By leveraging cloud platforms, cities can access scalable computing power without the need for expensive on-premises infrastructure. For example, Helsinki has used cloud computing to manage its Digital Twin for energy management, reducing infrastructure costs and allowing the city to scale its system over time (City of Helsinki, 2021). Similarly, edge computing enables cities to process data closer to where it is generated, improving real-time decision-making, and reducing latency. This approach has been successfully implemented in Copenhagen, where edge computing is used to optimize traffic flows and reduce congestion (City of Copenhagen, 2022b, c).

To address cybersecurity concerns, cities must invest in advanced security protocols, such as end-to-end encryption, AI-powered threat detection, and Zero Trust security models. These strategies can help protect sensitive data and ensure the integrity of Digital Twin systems. Singapore's Virtual Singapore project has implemented rigorous cybersecurity measures, including real-time monitoring and AI-based threat detection, to safeguard its Digital Twin against potential attacks (Government Technology Agency, 2021).

The development of global standards is also essential for overcoming the challenges related to interoperability and governance. International organizations such as the ISO and OGC are making progress in creating standards for data sharing and system integration. Cities that adopt these standards can ensure that their Digital Twin systems are interoperable with those in other regions, enabling greater collaboration and innovation (ISO, 2020a, b).

Finally, fostering public–private partnerships can provide cities with the resources and expertise needed to implement Digital Twin technology. By working with technology companies, academic institutions, and international organizations, cities can access cutting-edge solutions without shouldering the entire financial burden. In New York City, partnerships with companies like Google and IBM have enabled the development of a Digital Twin for managing public services, demonstrating the value of collaboration in overcoming financial and technical barriers (NYC Mayor's Office of Technology, 2020c, f).

By embracing cloud-based infrastructure, enhancing cybersecurity, establishing global standards, and fostering collaboration, cities can overcome the challenges of Digital Twin adoption. In doing so, they will pave the way for smarter, more sustainable cities capable of meeting the complex demands of the future.

REFERENCES

Amazon Web Services. (2021). *Digital twins on AWS: Powering smart cities through cloud technology*. Available at: https://aws.amazon.com [Accessed on 22 October 2024].

CISA. (2021). *Cybersecurity incident involving Colonial Pipeline ransomware attack*. Available at: https://www.cisa.gov [Accessed on 22 October 2024].

City of Amsterdam. (2021a). *Digital Twin City Project: Challenges in scaling and data processing*. Available at: https://www.amsterdam.nl [Accessed on 22 October 2024].
City of Amsterdam. (2021b). *Digital Twin City Project: Governance and interoperability challenges*. Available at: https://www.amsterdam.nl [Accessed on 22 October 2024].
City of Amsterdam. (2021c). *Digital twin technology and climate-neutrality by 2050*. Available at: https://www.amsterdam.nl [Accessed on 22 October 2024].
City of Barcelona. (2022). *Smart city data integration: Lessons from Barcelona's digital twin*. Available at: https://www.barcelona.cat [Accessed on 22 October 2024].
City of Copenhagen. (2022a). *Copenhagen's use of edge computing in real-time traffic management*. Available at: https://www.copenhagen.dk [Accessed on 22 October 2024].
City of Copenhagen. (2022b). *Cybersecurity and digital twin technology: Enhancing public transport safety*. Available at: https://www.copenhagen.dk [Accessed on 22 October 2024].
City of Copenhagen. (2022c). *Edge computing in Copenhagen's traffic management Digital Twin*. Available at: https://www.copenhagen.dk [Accessed on 22 October 2024].
City of Helsinki. (2021). *Cloud-based Digital Twin for energy management in Helsinki*. Available at: https://www.hel.fi [Accessed on 22 October 2024].
Deloitte. (2021). *5G infrastructure and its role in digital twin technology for smart cities*. Available at: https://www.deloitte.com [Accessed on 22 October 2024]
European Commission. (2021). *General Data Protection Regulation (GDPR) and data privacy in Digital Twins*. Available at: https://ec.europa.eu [Accessed on 22 October 2024].
Forrester. (2021). *Zero Trust security: Redefining cybersecurity for the digital age*. Available at: https://www.forrester.com [Accessed on 22 October 2024].
Gartner. (2022). *Cost analysis of cloud services for digital twin applications*. Available at: https://www.gartner.com [Accessed on October 2024]
Government Technology Agency. (2021). *Virtual Singapore: Shaping a Smart Nation with strong cybersecurity defenses*. Available at: https://www.tech.gov.sg [Accessed on 22 October 2024].
IIT Hyderabad. (2023). *Indo-Dutch partnership in smart city development and digital twin technologies*. Available at: https://www.iith.ac.in [Accessed on 22 October 2024].
ISO. (2020a). *Smart city standards for data interoperability and system integration*. Available at: https://www.iso.org [Accessed on 22 October 2024].
ISO. (2020b). *Smart city standards for governance and data interoperability: An ISO perspective*. Available at: https://www.iso.org [Accessed on 22 October 2024].
Kim, J. (2022). *Seoul's 5G-enabled digital twin: Optimizing public transport through real-time data. Journal of Smart City Development*, *9*(3), 45–60.
MarketsandMarkets. (2020). *Digital Twin Market – Global Forecast to 2026*. Available at: https://www.marketsandmarkets.com [Accessed on 22 October 2024]
McKinsey & Company. (2021). *The Future of Digital Twins: Challenges and Opportunities*. Available at: https://www.mckinsey.com [Accessed on 22 October 2024].
NYC Mayor's Office of Technology. (2020a). *Challenges in integrating legacy systems with digital twin technology: The New York case study*. Available at: https://www1.nyc.gov [Accessed on 22 October 2024].
NYC Mayor's Office of Technology. (2020b). *Data management and integration challenges in New York's Digital Twin project*. Available at: https://www1.nyc.gov [Accessed on 22 October 2024].
NYC Mayor's Office of Technology. (2020c). *Data privacy concerns in New York's digital twin pilot project*. Available at: https://www1.nyc.gov [Accessed on 22 October 2024].
NYC Mayor's Office of Technology. (2020d). *Integrating legacy infrastructure with digital twin systems: The New York case study*. Available at: https://www1.nyc.gov [Accessed on 22 October 2024].

NYC Mayor's Office of Technology. (2020e). *Public-private partnerships in Digital Twin technology: New York case study.* Available at: https://www1.nyc.gov [Accessed on 22 October 2024].

NYC Mayor's Office of Technology. (2020f). *Public-private partnerships in Digital Twin technology: Lessons from New York.* Available at: https://www1.nyc.gov [Accessed on 22 October 2024].

OGC. (2021). *Open standards for geospatial data sharing in smart cities.* Available at: https://www.ogc.org [Accessed on 22 October 2024].

Port of Rotterdam Authority. (2022). *Sustainability and digital twin technology: Reducing emissions in Europe's largest port.* Available at: https://www.portofrotterdam.com [Accessed on 22 October 2024].

RICS. (2021). *Cybersecurity and Digital Twins in Infrastructure.* Available at: https://www.rics.org [Accessed on 22 October 2024].

Sharma, L. (Ed.). (2020). Towards Smart World: Homes to Cities Using Internet of Things (1st ed.). Chapman and Hall/CRC. https://doi.org/10.1201/9781003056751

Sharma, L. (2022). "Preventing Security Breach in Social Media: Threats and Prevention Techniques", *Computer Vision and Internet of Things: Technologies and Applications",* Taylor & Francis, CRC Press, pp.53–62.

Sharma, L., & Carpenter, M. (Eds.). (2022). Computer Vision and Internet of Things: Technologies and Applications (1st ed.). Chapman and Hall/CRC. https://doi.org/10.1201/9781003244165

Transport for London. (2021). *Digital twin for transport: The future of London's Underground.* Available at: https://tfl.gov.uk [Accessed on 22 October 2024].

United Nations. (2021). *Urbanization and technology in low-income cities: The case of Dhaka, Bangladesh. United Nations Economic and Social Affairs.* Available at: https://www.un.org [Accessed on 22 October 2024].

Wang, X. (2021). Digital Twin technology and privacy concerns in China's smart cities. *Journal of Urban Technology*, *28*(2), 45–62.

World Bank. (2020). *The digital divide in smart city adoption: Addressing socioeconomic barriers.* Available at: https://www.worldbank.org [Accessed on 22 October 2024].

Zetter, K. (2017). *Inside the cunning, unprecedented hack of Ukraine's power grid. Wired.* Available at: https://www.wired.com [Accessed on 22 October 2024].